MICROBIOLOGY RESEARCH ADVANCES

STAPHYLOCOCCUS AUREUS AND COAGULASE-NEGATIVE STAPHYLOCOCCI

VIRULENCE, ANTIMICROBIAL RESISTANCE AND MOLECULAR EPIDEMIOLOGY

MICROBIOLOGY RESEARCH ADVANCES

Additional books in this series can be found on Nova's website under the Series tab.

Additional e-books in this series can be found on Nova's website under the e-book tab.

MICROBIOLOGY RESEARCH ADVANCES

STAPHYLOCOCCUS AUREUS AND COAGULASE-NEGATIVE STAPHYLOCOCCI

VIRULENCE, ANTIMICROBIAL RESISTANCE AND MOLECULAR EPIDEMIOLOGY

MARIA DE LOURDES RIBEIRO
DE SOUZA DA CUNHA

Copyright © 2014 by Nova Science Publishers, Inc.

All rights reserved. No part of this book may be reproduced, stored in a retrieval system or transmitted in any form or by any means: electronic, electrostatic, magnetic, tape, mechanical photocopying, recording or otherwise without the written permission of the Publisher.

For permission to use material from this book please contact us:
Telephone 631-231-7269; Fax 631-231-8175
Web Site: http://www.novapublishers.com

NOTICE TO THE READER

The Publisher has taken reasonable care in the preparation of this book, but makes no expressed or implied warranty of any kind and assumes no responsibility for any errors or omissions. No liability is assumed for incidental or consequential damages in connection with or arising out of information contained in this book. The Publisher shall not be liable for any special, consequential, or exemplary damages resulting, in whole or in part, from the readers' use of, or reliance upon, this material. Any parts of this book based on government reports are so indicated and copyright is claimed for those parts to the extent applicable to compilations of such works.

Independent verification should be sought for any data, advice or recommendations contained in this book. In addition, no responsibility is assumed by the publisher for any injury and/or damage to persons or property arising from any methods, products, instructions, ideas or otherwise contained in this publication.

This publication is designed to provide accurate and authoritative information with regard to the subject matter covered herein. It is sold with the clear understanding that the Publisher is not engaged in rendering legal or any other professional services. If legal or any other expert assistance is required, the services of a competent person should be sought. FROM A DECLARATION OF PARTICIPANTS JOINTLY ADOPTED BY A COMMITTEE OF THE AMERICAN BAR ASSOCIATION AND A COMMITTEE OF PUBLISHERS.

Additional color graphics may be available in the e-book version of this book.

LIBRARY OF CONGRESS CATALOGING-IN-PUBLICATION DATA

ISBN 978-1-63117-938-9 (softcover)

Library of Congress Control Number: 2014943983

Published by Nova Science Publishers, Inc. † New York

*"**God** wishes, man dreams, the work is born"*
Fernando Pessoa

I dedicate this book...
*To my parents, **Antenor** and **Alice***
*To my husband, **Toninho***

*And especially to my daughters, **Taís** and **Letícia***
*The most precious gift **God** allowed me to have*
Thank you for your love, for your constant presence,
Sharing with me my dreams, and for effectively contributing by drawing the illustrations of this book.

Contents

Acknowledgments		ix
List of Abbreviations and Acronyms		xi
Preface		xv
Chapter 1	The Genus *Staphylococcus*	1
Chapter 2	Clinical Significance of Coagulase-Negative Staphylococci	7
Chapter 3	Catheter-Related Infections	21
Chapter 4	Identification of *Staphylococcus* spp.	37
Chapter 5	Staphylococcal Biofilms	45
Chapter 6	Staphylococcal Toxins	61
Chapter 7	Antimicrobial Resistance	81
Chapter 8	Epidemiology of Methicillin-Resistant *Staphylococcus* spp.	95
Author's Contact Information		109
Index		111

ACKNOWLEDGMENTS

To *God*, for letting me see the right way, for permitting me to live with blessed people, and for guiding me.

To *Fundação de Amparo à Pesquisa do Estado de São Paulo (FAPESP)* grant 2011/09106-0, *Conselho Nacional de Desenvolvimento Científico e Tecnológico (CNPq)* and *Fundação para o Desenvolvimento da UNESP (FUNDUNESP)* for the financial support and fellowships that were fundamental for obtaining the results reported in this book.

To all my students who continuously encouraged me throughout all these years, too many to be cited individually, and who were and continue to be very important to me and without whom this book would not have been possible.

To *Prof. Dr. Carlos Magno Castelo Branco Fortaleza*, a great friend and collaborator with deep knowledge in the area of infectious diseases and epidemiology who read the preliminary text that gave origin to this book and contributed with his valuable suggestions and inestimable help.

To the team that participated directly in the execution of the studies included in this book: *Augusto Cezar Montelli, Adilson Oliveira, Adriano Martison Ferreira, André Martins, Alessandro Lia Mondelli, Ana Maria Fioravante, Camila Marconi, Camila Sena Martins de Souza, Carlos Alberto de Magalhães Lopes, Carlos Henrique Camargo, Carlos Magno Castelo Branco Fortaleza, Danilo Flávio Moraes Riboli, Deise Rafaela Ustulin, Eliane Peresi, Eliane Patricia Lino Pereira Franchi, Eliane Pessoa da Silva, Fabiana Venegas Pires, Jacqueline Teixeira Costa Caramori, Jackson Eliezer Neves Batalha, João Cesar Lyra, João Pessoa Araújo Júnior, José Eduardo Corrente, Letícia Teixeira Pazzini, Liciana Vaz de Arruda Silveira, Ligia Maria Suppo de Souza Rugolo, Lígia Maria Abraão, Marcus Vinicius Pimenta Rodrigues, Maria Fátima Sugizaki, Maria Regina Bentlin, Mariana Fávero*

Bonesso, Natalia Bibiana Teixeira, Nathalie Gaebler Vasconcelos, Pasqual Barretti, Patrícia Yoshida Faccioli Martins, Regina Adriana Oliveira Calsolari, Taíse Marongio Cotrim de Moraes, Valéria Cataneli Pereira.

To *Kerstin Markendorf* for her support and version of the book into English.

I am grateful for the privilege to live with so many intelligent and dedicated people that significantly contributed to my scientific career.

LIST OF ABBREVIATIONS AND ACRONYMS

Aap	Accumulation-associated protein
agr	Accessory gene regulator
AIP	Autoinducing peptide
ANVISA	Agência Nacional de Vigilância Sanitária (National Sanitary Surveillance Agency)
arl	Autolysis-related locus
ATCC	American Type Culture Collection
Atle	Autolysin
Bap	Biofilm-associated protein
Bhp	Bap-homologous protein
bp	Base pairs
CA-MRSA	Community-associated methicillin-resistant *Staphylococcus aureus*
CAPD	Continuous ambulatory peritoneal dialysis
CC	Clonal complex
CDC	Centers for Disease Control
cDNA	Complementary DNA
CFL	Cephalothin
CFO	Cefoxitin
CFU	Colony-forming units
CLSI	Clinical Laboratory Standards Institute
CoNS	Coagulase-negative staphylococci
CRA	Congo-red agar
CRBSI	Catheter-related bloodstream infection
CVC	Central venous catheter
DTP	Differential time to positivity

eDNA	Extracellular genomic DNA
egc	Enterotoxin gene cluster
ELISA	Enzyme-linked immunosorbent assay
ERY	Erythromycin
FAPESP	Fundação de Amparo à Pesquisa do Estado de São Paulo (São Paulo Research Foundation)
FMB	Faculdade de Medicina de Botucatu (Botucatu Medical School)
GEN	Gentamicin
GPI	Gram-positive identification
HACO-MRSA	Healthcare-associated community-onset methicillin-resistant *Staphylococcus aureus*
HAI	Healthcare-associated infection
HA-MRSA	Healthcare-associated methicillin-resistant *Staphylococcus aureus*
HC	Hospital das Clínicas da Faculdade de Medicina de Botucatu (University Hospital of the Botucatu Medical School)
ica	Intercellular adhesion locus
ICU	Intensive care unit
ID	Identification
ITS	Intergenic transcribed spacer
mgrA	Multiple global regulator
MIC	Minimum inhibitory concentration
MLST	Multilocus sequence typing
MSCRAMMs	Microbial surface components recognizing adhesive matrix molecules
MOD-SA	Modified oxacillin resistance
mRNA	Messenger RNA
MRSA	Methicillin-resistant *Staphylococcus aureus*
MSSA	Methicillin-sensitive *Staphylococcus aureus*
NHSN	National Healthcare Safety Network
NICU	Neonatal intensive care unit
NNIS	National Nosocomial Infection Surveillance System
OPT	Operon Technology
OXA	Oxacillin
PBP	Penicillin-binding protein
PCR	Polymerase chain reaction
PEN	Penicillin

PFGE	Pulsed-field gel electrophoresis
PIA	Polysaccharide intercellular adhesin
PMNs	Polymorphonuclear leukocytes
PPV	Positive predictive value
PSM	Phenol-soluble modulin
PTSAg	Pyrogenic toxin superantigen
PVL	Panton-Valentine leukocidin
QS	Quorum sensing
RAP	RNAIII-activating protein
RAPD-PCR	Random amplified polymorphic DNA-PCR
REP-PCR	Repetitive extragenic palindromic sequence-based PCR
RIF	Rifampicin
RIP	RNAIII-inhibiting peptide
ROC	Receiver operating characteristic
rot	Repressor of toxins
RPLA	Reverse passive latex agglutination
RT-PCR	Reverse transcriptase-polymerase chain reaction
sae	Staphylococcal accessory element
SaPI	*Staphylococcus aureus* pathogenicity island
sar	Staphylococcal accessory regulator
SCC*mec*	Staphylococcal cassette chromosome *mec*
SCOPE	Surveillance and Control of Pathogens of Epidemiological Importance
SEA	Staphylococcal enterotoxin A
SEB	Staphylococcal enterotoxin B
SEC	Staphylococcal enterotoxin C
sec-1	Staphylococcal enterotoxin gene C subtype 1
SEC$_3$	Staphylococcal enterotoxin C subtype 3
SED	Staphylococcal enterotoxin D
SEE	Staphylococcal enterotoxin E
SEG	Staphylococcal enterotoxin G
SEH	Staphylococcal enterotoxin H
SEI	Staphylococcal enterotoxin I
SEl	Staphylococcal enterotoxin-like
SElJ	Staphylococcal enterotoxin-like J
SElK	Staphylococcal enterotoxin-like K
SElM	Staphylococcal enterotoxin-like M
SElN	Staphylococcal enterotoxin-like N
SElO	Staphylococcal enterotoxin-like O

SElP	Staphylococcal enterotoxin-like P
SElQ	Staphylococcal enterotoxin-like Q
SePI	*Staphylococcus epidermidis* pathogenicity island
SER	Staphylococcal enterotoxin R
SES	Staphylococcal enterotoxin S
SET	Staphylococcal enterotoxin T
spa typing	Protein A gene typing
srr	Staphylococcal respiratory response
SSR	Simple sequence repeat
ST	Sequence type
TRAP	Target protein of RAP
TSS	Toxic shock syndrome
TSST-1	Toxic shock syndrome toxin 1
UNESP	Universidade Estadual Paulista (Paulista State University)
UNIFESP	Universidade Federal do Estado de São Paulo (Federal University of São Paulo)
USA	United States of America
UTI	Urinary tract infection
VAN	Vancomycin
VISA	Vancomycin-intermediate *Staphylococcus aureus*
VRE	Vancomycin-resistant enterococci
VRSA	Vancomycin-resistant *Staphylococcus aureus*

PREFACE

In this book, I decided to discuss studies related to my line of research "*Staphylococcus aureus* and coagulase-negative staphylococci: virulence, antimicrobial resistance and molecular epidemiology". I became interested in this line of research already as a scientific initiation student during my undergraduate course in Biological Sciences at the State University of Londrina, PR, Brazil, when I began to study the detection of staphylococcal enterotoxins in *S. aureus* under the supervision of Prof. Dr. Elisa Yoko Hirooka. During my PhD, already as a professor of the Department of Microbiology and Immunology at the Institute of Biosciences (IB), UNESP, Botucatu, SP, Brazil, under the supervision of Prof. Dr. Carlos Alberto de Magalhães Lopes, I began to work with other species of the genus *Staphylococcus*, the coagulase-negative staphylococci (CoNS). My research focused on the identification of these species, as well as the phenotypic detection of different virulence factors, including enzymes, biofilm, enterotoxins and antimicrobial resistance in strains isolated from newborns of the Neonatal Unit of the University Hospital (HC), Botucatu Medical School (FMB), SP, Brazil, in collaboration with Prof. Dr. Ligia Maria S. S. Rugolo and, subsequently, with the neonatologists Prof. Dr. João César Lyra and Prof. Dr. Maria Regina Bentlin, researchers at the Department of Pediatrics, FMB. After my PhD, I was invited by Dr. Augusto Cezar Montelli to join the research group on peritonitis in peritoneal dialysis, together with Prof. Dr. Pasqual Barretti and Prof. Dr. Jacqueline Teixeira Caramori, the researchers responsible for the Dialysis Unit of HC-FMB, Botucatu, SP, Brazil.

In view of the results of my PhD and of the difficulty in convincing the reviewers of the truthfulness of the results obtained regarding the toxigenicity of CoNS and the need to confirm the results using more reliable genotypic methods, I entered the area of Molecular Biology in collaboration with Prof. Dr. João Pessoa Araújo Júnior of the Department of Microbiology and Immunology, IB, UNESP, Botucatu, SP, Brazil. In parallel to these projects on pathogenicity factors, I began to study the detection of oxacillin resistance by molecular methods. Thereafter, I entered the area of Molecular Epidemiology in collaboration with Prof. Dr. Carlos Magno Castelo Branco Fortaleza of the Department of Tropical Diseases, FMB, UNESP, Botucatu, Brazil.

The book covers updated topics on the genus *Staphylococcus*, including the latest discoveries. In each chapter I tried to discuss the results obtained and published during my more than 20 years as a researcher in this area.

The book is divided into eight chapters. The first chapter is a presentation of the genus *Staphylococcus* and focuses on the current classification, the species described, and the general characteristics of these bacteria. The second chapter discusses the clinical significance of CoNS, which are often classified as mere blood culture contaminants, but are the etiological agents most commonly associated with neonatal infections and peritonitis in patients undergoing peritoneal dialysis. These infections occur mainly in patients submitted to invasive procedures. Catheter-related infections are an example of this scenario. This topic is discussed in chapter 3, with emphasis on diagnostic techniques.

Chapter 4 focuses on the identification of CoNS, which is important for predicting the pathogenic potential and antibiotic susceptibility profile of each species, thus permitting more adequate assessment of the clinical significance of specific species.

Coagulase-negative staphylococci can colonize the surface of catheters. Once adhered, these bacteria proliferate, forming multiple layers, and produce an extracellular polysaccharide that leads to the formation of biofilms which potentiate their pathogenicity. This topic is discussed in chapter 5.

Chapter 6 discusses the virulence factors responsible for the symptoms and severity of infections caused by *Staphylococcus* spp. These factors include membrane-active toxins, called cytotoxins, and a group of pyrogenic toxin superantigens. In this chapter, I discuss several current aspects related to the virulence of *S. aureus* and CoNS, including isolates of different origins, phenotypic and genotypic techniques for the detection of these toxins, and the gene regulation mechanisms involved in their expression.

Another extremely important topic, which is discussed in chapter 7, is the resistance of these microorganisms to antimicrobial agents. Half a century after the description of the first isolates of methicillin-resistant *Staphylococcus aureus* (MRSA) in England, these microorganisms have spread around the world. Although much has been learned about these bacteria, we have been unable to fully eradicate them or to consistently prevent the severe infections caused by these microorganisms. MRSA currently represent a serious threat to public health worldwide due to the rapid dissemination and diversification of pandemic clones that show increasing virulence and antimicrobial resistance. These microorganisms are one of the main causes of healthcare-associated infections and the prevalence of community-associated methicillin-resistant *Staphylococcus aureus* (CA-MRSA) has also been increasing. The epidemiology of MRSA infections is discussed in chapter 8. This chapter highlights the importance of understanding why specific clones predominate in different regions and of identifying individual factors related to their acquisition in order to develop more effective control strategies and to choose more appropriate antimicrobial therapy.

In this book, the topics are explained in a clear and objective manner and should be useful for postgraduate students, physicians and researchers who need to better understand this fascinating group of microorganisms.

Chapter 1

THE GENUS *STAPHYLOCOCCUS*

The name *Staphylococcus* is derived from greek terms (staphyle = bunch of grapes, and coccus = grain or seed) and was given by Alexander Ogston, a Scottish surgeon, in 1880 when he isolated this microorganism from a surgical abscess, referring to the morphology and arrangement of these bacteria observed under a light microscope. Performing various experiments, he was able to demonstrate the importance of staphylococci in purulent human infections [1]. However, it was only in 1884 that the researcher Rosenbach, who studied microorganisms isolated from pus, proposed the genus *Staphylococcus*. Since its proposal by Rosenbach, the genus *Staphylococcus* has been classified within the family Micrococcaceae and it was only in the last decade with the advances in molecular biology and genetics and studies of fatty acid profiles and cell wall composition, and particularly studies on 16S ribosomal RNA, that the genus *Staphylococcus* was included in the new family Staphylococcaceae [2, 3]. In contrast to Micrococcaceae, the family Staphylococcaceae belongs to the phylum Firmicutes, class Bacilli, order Bacillales, with the demonstration of the phylogenetic distance between the genera *Staphylococcus* and *Micrococcus* more than a century after its first description and of the importance of advances in molecular biology for the correct taxonomic classification of these microorganisms.

Staphylococci are Gram-positive, immobile, facultative anaerobic, non-photosynthetic, non-sporulating and catalase-positive cocci that produce acid, but no gas, by fermentation and are able to grow in medium containing 10% sodium chloride. They are mesophilic microorganisms that grow at a temperature range of 7 to 48°C (optimum temperature: 37°C) and pH of 4.0 to 10.0 (optimum pH: 6.0 to 7.0) [4].

Staphylococci grow rapidly in most bacteriological media and produce round, smooth, elevated and bright colonies on solid media. These bacteria synthesize a carotenoid pigment only under aerobic conditions, which is intensified in media containing high salt concentrations. The ability of staphylococci to grow in high-NaCl media and a certain tolerance to potassium tellurite are utilized for the preparation of selective media.

The genus *Staphylococcus* comprises different species and subspecies which are widely distributed in nature and are mainly found on the skin and mucous membranes of birds and mammals [4]. Forty-nine species of the genus have been described so far [5, 6], most of them coagulase-negative. The enzyme coagulase is synthesized exclusively by the species *S. aureus* subsp. *aureus*, *S. aureus* subsp. *anaerobius*, *S. hyicus*, *S. intermedius*, *S. pseudintermedius*, *S. schleiferi* subsp. *coagulans*, *S. delphini*, *S. lutrae*, and *S. agnetis*. *Staphylococcus hyicus* and *S. agnetis* are variably coagulase positive and are frequently included among coagulase-negative microorganisms [5, 6]. About half of coagulase-negative staphylococci (CoNS) naturally colonize humans, including *S. epidermidis*, *S. haemolyticus*, *S. saprophyticus*, *S. lugdunensis* [7], *S. xylosus* [8], *S. warneri*, *S. simulans* [9], *S. saccharolyticus* [10], *S. auricularis* [11], *S. caprae* [12], *S. pasteuri* [13], *S. vitulinus* [14], *S. pettenkoferi* [15], *S. massiliensis* [16], and *S. jettensis*, a species described recently[17]. Among the 26 CoNS subspecies described, *S. hominis* subsp. *hominis*, *S. hominis* subsp. *novobiosepticus*, *S. capitis* subsp. *capitis*, *S. capitis* subsp. *urealyticus* [18], *S. schleiferi* subsp. *schleiferi*, *S. cohnii* subsp. *cohnii*, *S. cohnii* subsp. *urealyticum* [19], *S. petrasii* subsp. *petrasii*, and *S. petrasii* subsp. *croceilyticus* [20] are also part of the normal bacterial flora of humans and other primates.

Staphylococcus aureus has always been the most important species which is associated with a series of infections and food poisoning in humans and animals. Several virulence factors are responsible for the symptoms and severity of community- and hospital-acquired *S. aureus* infections, which are currently a leading clinical and epidemiological problem in healthcare-associated infections. *Staphylococcus aureus* is important because of its pathogenicity and high frequency. This bacterium can cause diseases in both immunocompromised and healthy individuals due to its easy intrahospital dissemination and enormous adaptation capacity and resistance to antimicrobial drugs [21].

CoNS are the most common members of the human microbiota [22] and can reach 10^3 to 10^6 colony-forming units (CFU)/cm^2 on the surface of moist areas of the body, such as the anterior nares, axillae and inguinal and perineal areas. Some species and subspecies have a marked preference for certain habitats; for example, *S. capitis* subsp. *capitis* is found in large numbers on the head, forehead, eyebrows and external auditory canal [22]. In contrast, *S. capitis* subsp. *urealyticus* is less frequent at these sites, but is widely distributed in other parts of the body [5]. *Staphylococcus auricularis* is one of the most common species inhabiting the external auditory canal of humans [11], whereas *S. saprophyticus* is generally found in small, transient populations at a variety of body sites, but specifically adheres to urogenital cells [23]. *Staphylococcus epidermidis* is the predominant species in humans and is found in large numbers in the anterior nares, axillae and perineal area [24]. *Staphylococcus hominis* and *S. haemolyticus* are also numerous in moist places of the body, but colonize drier areas of the skin more frequently than other species [24]. *Staphylococcus warneri* and *S. lugdunensis* are found throughout the body, but their numbers are generally low [5]. *Staphylococcus caprae* and *S. xylosus*, although typically found in nature, are occasionally isolated from human skin [9].

However, CoNS are also considered opportunistic microorganisms which can take advantage of some situations, such as disruption of the skin barrier due to trauma or the presence of foreign bodies, and can thus spread to other tissues, proliferate and develop a pathogenic behavior [25]. Usually, CoNS cause infection in immunocompromised patients or patients with catheters. The main reasons for the increase in the rate of infections caused by CoNS in recent years is the growing resistance to antimicrobial drugs among these bacteria and the increasing use of medical devices [26].

REFERENCES

[1] Holt, JG; Krieg, NR; Sneathm, PHA; Staley, JT; Williams, ST. *Manual of determinative bacteriology*. 9th ed. Baltimore, MD: Williams and Wilkins, 1994.

[2] Schleifer, KH; Bell; JA., Family, VIII. Staphylococcaceae fam. nov. In: Vos, P, Garrity GM, Jones D, Krieg NR, Ludwig W, Rainey FA, Schleifer KH, Whitman, HB, editors. *Bergey's Manual of Systematic Bacteriology*. New York: Springer, 2009, p. 392.

[3] Euzéby, J. List of new names and new combinations previously effectively, but not validly, published. *Int J Syst Evol Microbiol.*, 2010, 60(Pt 5), 1009-10

[4] Kloos, WE; Bannerman, TL. *Staphylococcus* and *Micrococcus*. In: Murray PR, Baron EJ, Pfaller MA, Tenover FC, Yolken RH, editors. Manual of Clinical Microbiology. Washington: American Society for Microbiology Press, 1999, 264-282.

[5] Bannerman, TL. *Staphylococcus, Micrococcus*, and other catalase-positive cocci that grow aerobically. In: Murray PR, Baron EJ, Jorgensen JH, Pfaller MA, Yolken RH, editors. Manual of Clinical Microbiology. *Washington: American Society for Microbiology*, 2003, p. 384-404.

[6] Euzéby; JP. List of prokaryotic names with standing in nomenclature – Genus *Staphylococcus* [Internet]. Accessed March 10, 2014. Available at: http://www.bacterio.cict.fr/s/staphylococcus.html.

[7] Freney, J; Brun, Y; Bes, M; Meugnier, H; Grimont, F; Grimont, PAD; Nervi, C; Fleurette, J. *Staphylococcus lugdunensis* sp. nov. and *Staphylococcus schleiferi* sp. nov., two species from human clinical specimens. *Int J Syst Bacteriol.*, 1988, 38(2),168-172.

[8] Schleifer, KH; Kloos, WE. A simple test system for the separation of staphylococci from micrococci. *J Clin Microbiol.*, 1975, 1(3), 337-8.

[9] Kloos, WE; Schleifer, KH. Simplified scheme for routine identification of human *Staphylococcus* species. *J Clin Microbiol.*, 1975, 1(1), 82-8.

[10] Westblom; TU; Gorse, GJ; Milligan, TW; Schindzielorz. Anaerobic endocarditis caused by *Staphylococcus saccharolyticus*. *J Clin Microbiol.*, 1990, 28(12), 2818-2819.

[11] Kloos, WE; Schleifer, KH. *Staphylococcus auricularis* sp. nov.: an inhabitant of the human external ear. *Int J Syst Bacteriol.*, 1983, 33(1),9-14.

[12] Devriese, LA; Poutrel, B; Kilpper-Balz, R; Schleifer, KH. *Staphylococcus gallinarum* and *Staphylococcus caprae*, two new species from animals. *Int J Syst Bacteriol.*, 1983, 33(3),480-486.

[13] Chesneau, O; Morvan, A; Grimont, F; Labischinski, H; el, Solh, N. *Staphylococcus pasteuri* sp. nov., isolated from human, animal, and food specimens. *Int J Syst Bacteriol.*, 1993, 43(2), 237-44.

[14] Svec, P; Vancanneyt, M; Sedlácek, I; Engelbeen, K; Stetina, V; Swings, J; et al. Reclassification of *Staphylococcus pulvereri* Zakrzewska-Czerwinska et al. 1995 as a later synonym of *Staphylococcus vitulinus* Webster et al. 1994. *Int J Syst Evol Microbiol.*, 2004, 54(Pt 6), 2213-5.

[15] Trülzsch, K; Grabein, B; Schumann, P; Mellmann, A; Antonenka, U; Heesemann, J; et al. *Staphylococcus pettenkoferi* sp. nov., a novel

coagulase-negative staphylococcal species isolated from human clinical specimens. *Int J Syst Evol Microbiol.*, 2007, 57(Pt 7), 1543-8.
[16] Al, Masalma, M; Raoult, D; Roux, V. *Staphylococcus massiliensis* sp. nov., isolated from a human brain abscess. *Int J Syst Evol Microbiol.*, 2010, 60(Pt 5), 1066-72.
[17] De Bel, A; Van, Hoorde, K; Wybo, I; Vandoorslaer, K; Echahidi, F; De, Brandt, E; et al. *Staphylococcus jettensis* sp. nov., a coagulase-negative staphylococcal species isolated from human clinical specimens. *Int J Syst Evol Microbiol*, 2013, 63(Pt 9),3250-6.
[18] Bannerman, TL; Kloos, WE. *Staphylococcus capitis* subsp. *ureolyticus* subsp. nov. from human skin. *Int J Syst Bacteriol*. 1991, 41(1), 144-7.
[19] Kloos, WE; Wolfshohl, JF. *Staphylococcuscohnii* subspecies: *Staphylococcus cohnii* subsp. *cohnii* subsp. nov. and *Staphylococcus cohnii* subsp. *urealyticum* subsp. nov. *Int J Syst Bacteriol.*, 1991, 41(2), 284-9.
[20] Pantůček, R; Švec, P; Dajcs, Jj; Machová, I; Černohlávková, J; Šedo, O; et al. *Staphylococcus petrasii* sp. nov. including *S. petrasii* subsp. *petrasii* subsp. nov. and *S. petrasii* subsp. *croceilyticus* subsp. nov., isolated from human clinical specimens and human ear infections. *Syst Appl Microbiol.*, 2013, 36(2), 90-5.
[21] Enright, MC; Robinson, DA; Randle, G; Feil, EJ; Grundmann, H; Spratt, BG. The evolutionary history of methicillin-resistant *Staphylococcus aureus* (MRSA). *Proc Natl Acad Sci U S A*. 2002, 99(11), 7687-92.
[22] Kloos, WE; Bannerman, TL., *Staphylococcus* and *Micrococcus*. In: Murray, PR; Baron, EJ; Pfaller, MA; et al. editors. Manual of Clinical Microbiology. Washington: American Society for Microbiology, 1995. 282-298.
[23] Colleen, S, Hovelius, B, Wieslander, A; Mårdh, PA. Surface properties of *Staphylococcus saprophyticus* and *Staphylococcus epidermidis* as studied by adherence tests and two-polymer, aqueous phase systems. *Acta Pathol Microbiol Scand B.*, 1979, 87(6), 321-8.
[24] Kloos, WE; Musselwhite, MS. Distribution and persistence of *Staphylococcus* and *Micrococcus* species and other aerobic bacteria on human skin. *Appl Microbiol.*, 1975, 30(3), 381-5.
[25] Koneman, EW; Allen, SD; Janda, WM; Schreckenberger, PC. Color Atlas and Textbook of Diagnostic Microbiology. 5th ed. Philadelphia: J.B. Lippincott, 1997.
[26] Otto, M. Virulence factors of the coagulase-negative *Staphylococci*. *Front Biosci.*, 2004, 9, 841-63.

Chapter 2

CLINICAL SIGNIFICANCE OF COAGULASE-NEGATIVE STAPHYLOCOCCI

Until the 1970s, few reports of infections caused by CoNS had been published and these microorganisms were recognized by clinicians and microbiologists as contaminants of clinical samples, with *S. aureus* being the only pathogenic species within the genus *Staphylococcus* [1]. However, this distinction, which is frequently used for the clinical diagnosis, has been considered a challenge in terms of the role that these microorganisms play in infectious diseases.

The first two reports in the literature associating CoNS with pathogenic processes date back to 1945 when Herbst and Merricks [2] described a case of septicemia associated with *S. albus* in a patient submitted to percutaneous removal of a kidney stone. In the following year, Harry [3] reported a case of neonatal meningitis caused by the same species. *Staphylococcus albus* was subsequently renamed *S. epidermidis*. Thirteen years later, Smith et al. [4] described the involvement of this microorganism in three cases of septicemia. Several years later, Pulverer and Halswick [5] reported 128 cases of endocarditis caused by CoNS. In 1985, at the Fifth International Symposium on Staphylococci and Staphylococcal Infections, Pulverer [6] shared his frustration with the medical community regarding the difficulty in convincing the editors of a German medical journal that their data were serious when trying to publish their work in 1967. Wilson and Stuart [7] detected CoNS in 53 pure cultures of 1,200 cases of wound infections (4.4%). Pulverer and Pillich [8] published the incidence of CoNS in pyogenic infections in Germany, presenting data for 1960, 1969 and 1970. CoNS were detected in about 10% of pyogenic lesions. In 1962, Pereira [9] reported that *S.*

saprophyticus can cause urinary tract infection (UTI). A few years later, Gallagher et al. [10] and Mabeck [11] also described cases of UTI caused by *S. saprophyticus*.

In many laboratories, *S. epidermidis* was used as a generic and collective term for all other CoNS. This situation changed radically in the 1980s when the National Nosocomial Infections Surveillance System (NNIS) of the Centers for Disease Control and Prevention (CDC) demonstrated a significant increase in nosocomial infections caused by different CoNS species [12].

Considerable progress in the systematic classification of staphylococci and in the development of methods for the identification of the genus, species and subspecies has permitted clinicians to become aware of the variety of CoNS present in clinical samples, and hence to consider them etiological agents of a series of infectious diseases [1]. Today, CoNS are recognized as essentially opportunistic microorganisms that take advantage of numerous organic situations to produce severe infections. Data from a surveillance study conducted in the United States (Surveillance and Control of Pathogens of Epidemiological Importance, SCOPE) comprising a period of 7 years (March 1995 to September 2002) indicated Gram-positive microorganisms as the main causative agents of bloodstream infections (65%) and CoNS as the most frequent agents (31%), followed by *S. aureus* (20%) [13]. A recent (June 2007 to March 2010) Brazilian multicenter study using the same methodology as the SCOPE program of the United States, Brazilian SCOPE, investigated the epidemiology and microbiology of nosocomial bloodstream infections in patients from 16 Brazilian hospitals of different sizes and locations. In that study, *S. aureus* (14%) and CoNS (12.6%) were the most commonly isolated microorganisms [14].

Data reported by the National Healthcare Safety Network (NHSN) of the CDC, comprising the period from January 2006 to October 2007, ranked CoNS first place and *S. aureus* second place in the etiology of healthcare-associated infection (HAI) [15]. In Brazil, data from the Surveillance System for Hospital Infections in the State of São Paulo, Center of Epidemiological Surveillance, showed CoNS to be the most common agents associated with bloodstream infections in 2010, with a predominance of *S. epidermidis* and other CoNS (30.1%), followed by *S. aureus* (16.6%). These rates were similar to those observed in 2009 [16]. According to recent data from the NHSN (January 2009 to December 2010), *S. aureus* occupied first place and CoNS third place in the etiology of HAI [17]. However, when HAI types were analyzed separately, CoNS continued to occupy first place in the etiology of bloodstream infections.

CoNS maintain a symbiontic or commensal relationship with their hosts. In some situations, they develop mechanisms to interfere with the growth of other pathogenic bacteria, suggesting co-evolution between the microorganism and its host [18]. *Staphylococcus epidermidis* has only been described as a commensal microorganism, but its importance as an opportunistic pathogen is now recognized, especially because this species is one of the most common causative agents of nosocomial infections[19]. Infections caused by *S. epidermidis* are not as severe as those due to *S. aureus* since the former does not possess many virulence factors, affecting mainly preterm infants, immunosuppressed patients, and individuals with prosthetic implants [20]. The low level of virulence maintained by *S. epidermidis* may be due to the adaptive process in the host-pathogen system. Evolution favors species that cause little or no damage to their host and, in the case of less virulent microorganisms, these adaptations offer evolutionary advantages by permitting the prolonged duration of infection and possible transmission from one host to another[21].

Massey et al. [21] developed a mathematical model to understand why it is advantageous for *S. epidermidis* to maintain a low level of virulence compared to *S. aureus*. This model considered the colonization of the entire epithelium with *S. epidermidis* in all humans, whereas *S. aureus* almost exclusively colonizes the nares of some individuals, and also factors of gene regulation involved in colonization and bacterial interference. According to this model, *S. epidermidis* is spread more easily than *S. aureus* and the low virulence is associated with the potential of the microorganism to secrete factors that promote its persistence in the host rather than an aggressive attack [19, 21].

CoNS are the main cause of bacteremia among patients in intensive care units (ICUs) and neonatal intensive care units (NICUs) [1], especially among low birth weight infants who are immunologically immature and frequently require invasive procedures for the administration of nutrition and medications [22]. The increase in the incidence of nosocomial bacteremia caused by CoNS among newborns in the last 20 years has also been associated with the increased survival of low birth weight preterm infants (less than 1,500 g) and their prolonged hospital stay [22, 23].

However, the interpretation of blood cultures that are positive for CoNS is particularly difficult since these microorganisms colonize the skin and mucous membranes as commensals and may contaminate blood cultures during sample collection [1, 24]. Therefore, researchers have used a variety of clinical and laboratory criteria to distinguish contamination from bacteremia. In this respect, the diagnosis of bacteremia is made based on clinical data of the patients and isolation of identical microorganisms from two or more blood cultures. Cultures in which multiple CoNS strains or species grow in association with other microorganisms are classified as contaminants [1, 25, 26]. However, since the blood volume that can be obtained from low birth weight preterm infants is small, only one blood culture is generally performed to prevent the need for and risks of transfusions due to constant venipuncture [1, 27]. As a consequence, neonatologists have relied on clinical and laboratory criteria such as thermal instability, bradycardia, apnea, food intolerance, worsening of respiratory distress, glucose intolerance, hemodynamic instability, and hypoactivity/lethargy [25, 28].

D'Angio et al. [29] reported a rate of colonization with CoNS of 50% to 80% up to 4 days after the admission of newborns to the NICU and observed an increase in multidrug resistance from 32% to 82% at the end of one week in the NICU. There is no doubt that the isolation of CoNS from blood, cerebrospinal fluid and urine samples of a newborn with signs and symptoms of sepsis is significant, but may frequently represent contamination at the time of collection. According to criteria of the CDC [26, 28] and the Brazilian National Sanitary Surveillance Agency (Agência Nacional de Vigilância Sanitária - ANVISA) [25], CoNS should be isolated from at least two blood cultures collected at two different sites at a maximum interval of 48 hours between samplings. In the case of isolation of CoNS from only one blood culture, the clinical evolution of the patient, complementary tests (blood count and C-reactive protein), and growth of the microorganism in the first 48 hours of incubation should be taken into consideration. Growth after this period suggests contamination. In addition, if the positive sample was collected only from a central venous catheter (CVC), CoNS should not be considered the etiological agent of infection.

Among CoNS species found in humans, *S. epidermidis* is clinically the most important for newborn infants. In the United States, this bacterium is responsible for 10% to 27% of all cases of sepsis that occur in NICUs, with rates of 55% in very low birth weight infants ($\leq 1,500$ g) [30]. The main manifestations reported in a study of sepsis in preterm infants were apnea and bradycardia (88%) and the need for oxygen (59%) and mechanical ventilation (69%), whereas laboratory markers of the acute phase were poorly sensitive [31]. The clinical presentation includes sepsis, meningitis with or without cerebrospinal fluid alterations, necrotizing enterocolitis, pneumonia, omphalitis, soft tissue abscesses, endocarditis, abscesses, and osteomyelitis at the sites of venipuncture. Lethality is low, in agreement with the results of a study of newborn infants conducted in the Neonatal Unit of the University Hospital of the Botucatu Medical School (Hospital das Clínicas da Faculdade de Medicina de Botucatu – HC-FMB), Botucatu, State of São Paulo, Brazil, which showed a lethality rate of 13% of infections caused by CoNS [32]. In that study, 117 CoNS strains were isolated from 107 newborns; of these, 51% were considered pathogenic and 49% contaminant. Most of the infected infants were born premature (80%), half of them very low birth weight, i.e., immunologically immature. In addition, infection was associated with two or more invasive procedures, with 89% of the infants having a CVC, 65% receiving parenteral nutrition, and 61% mechanical ventilation. Multivariate regression analysis revealed that birth weight < 1,500 g increased the risk of infection by 6 times, the presence of a foreign body increased the risk by 4.4 times, and previous antibiotic therapy increased the risk by 5.4 times. The most commonly isolated species was *S. epidermidis* (78% of cases), which was identified in 87% of infections and in 65% of contaminations. Also in Brazil, Silbert et al. [33] evaluated the prevalence of infection versus contamination in patients younger than 60 days with CoNS-positive blood cultures and reported that only 27% of 41 newborns with positive blood cultures were considered to be infected, whereas the remaining 73% were considered to be cases of contamination or doubtful cases. Therefore, the major difficulty in the diagnosis of CoNS infection is contamination at the time of material collection, which is significant for blood samples and even worse for samples collected from foreign bodies and secretions as demonstrated in the studies of Cunha et al.[32] and Silbert et al. [33].

Other CoNS species, including two *S. haemolyticus* strains, three *S. lugdunensis* strains, one *S. simulans* strain, one *S. warneri* strain and one *S. xylosus* strains, were also isolated from children with clinical signs of pneumonia, necrotizing enterocolitis, and sepsis [32]. The identification of CoNS species is a useful marker of infection since *S. epidermidis* is the etiological agent most frequently associated with infectious diseases [32]. Silbert et al.[33] identified 91.1% of isolates as *S. epidermidis*, three (6.7%) as *S. hominis* and one (2.2%) as *S. warneri*. *Staphylococcus epidermidis* was only detected in the group of infected patients. These results confirm the findings of Lowy and Hammer[34] who believe that the identification of CoNS species is important to differentiate between contamination and infection.

Multicenter studies have shown the importance of CoNS in neonatal infections. In Australia, a 10-year study involving 18 neonatal units identified CoNS as the etiological agent of 1,281 cases of sepsis, corresponding to 57.1%

of all episodes of late-onset sepsis, with an incidence of 3.46/1,000 live births. Most cases (71%) were preterm infants born at 24-29 weeks of gestation [35]. Recently, similar results have been reported by Vergnano et al. [36] in a multicenter study conducted in England involving 12 neonatal units, with 358 episodes of sepsis caused by CoNS in 321 children. The authors observed a total incidence of infection of 3/1,000 live births when other etiologies were considered and an increase to 8/1,000 live birth when episodes caused by CoNS were analyzed. Most CoNS infections occurred in children ≤ 32 weeks of gestation (86%) and extremely low birth weight infants (65%).

Staphylococcus spp. are also the most common etiological agents of continuous ambulatory peritoneal dialysis (CAPD)-related peritonitis in the world. Bacterial peritonitis continues to be the most severe complication of peritoneal dialysis and is the most frequent cause of discontinuation of dialysis [37], with an important impact on mortality [38]. The clinical presentation and progression of peritonitis episodes are strongly influenced by the characteristics of the causative agent. CoNS are the most common etiological agents [37], whereas *S. aureus* is associated with severe infections and with high lethality.

The incidence of peritonitis caused by *S. aureus* in the Dialysis Unit of HC-FMB decreased significantly from 0.13 episodes/patient/year between 1996 and 2000 to 0.04 episodes/patient/year between 2006 and 2010 ($p = 0.03$) [39]. Although there are no data that could explain the change in the prevalent etiology, one may suggest that the implementation of care routines designed to prevent catheter-related infections and to eradicate *S. aureus* from nasal carriers has contributed to the reduction in peritoneal infections caused by this microorganism. However, episodes of peritonitis due to *S. aureus* can lead to persistent infection, higher risk of hospitalization, catheter removal, and death [37-39].

Peritonitis caused by CoNS is cured in most cases without other complications [40]; however, recurrence of apparently cured infections is frequently observed. Few studies have compared the clinical course of infections caused by these species in recent years [41]. In this respect, it would be important to determine whether the progressively increasing resistance rates observed among CoNS had an impact on the evolution and complications of infections caused by these microorganisms. Our research group has therefore conducted several studies involving patients undergoing CAPD at the Dialysis Unit of HC-FMB in order to describe the microbiological properties of staphylococci causing peritonitis, to compare infections due to *S. aureus* with those caused by CoNS, and to establish associations between characteristics of the pathogen and host in the evolution of these infections.

Analysis of peritonitis episodes that occurred between January 1996 and December 2000 showed that the resolution of peritonitis was not influenced by host factors (age, gender, diabetes, vancomycin use, exchange system, or duration of dialysis), whereas *S. aureus* etiology was independently associated with the lack of resolution when compared to peritonitis caused by CoNS [42]. Factors related to the species and antimicrobial resistance could explain these results. However, no difference in the rate of resolution was observed between oxacillin-resistant and -susceptible strains. Since oxacillin resistance is more frequent among CoNS than in *S. aureus*, the contribution of antimicrobial resistance is inconsistent. The results of that study suggested virulence factors more frequently found in *S. aureus* to be responsible for the more aggressive nature of *S. aureus* and the consequent poorer outcome.

Another study published by our group [43] using two logistic regression models supported our hypothesis. In the first model that did not include virulence factors, the probability of resolution was not influenced by host factors, but was higher for episodes caused by *S. epidermidis* compared to *S. aureus* (p = 0.0263). In contrast, using the second model that included virulence factors, no difference in the probability of peritonitis resolution was observed between episodes caused by *S. aureus* and *S. epidermidis*. This finding may be explained by the fact that the inclusion of enzymes and toxins in the model permitted to control for the effect of these factors on the species, i.e., the species effect observed in the first model for *S. aureus* episodes may be due to the effect of pathogenicity factors that are more frequent in this species.

Furthermore, in the second model the episodes caused by *S. epidermidis* presented a lower rate of resolution than those caused by other CoNS species, irrespective of virulence factors. These results highlight the importance of the identification of CoNS species, which indeed behave differently and should not be evaluated as a group, but rather as distinct species with specific characteristics. Recent studies describe *S. epidermidis* as a versatile microorganism that can live as a commensal and pathogenic bacterium. In addition, this species employs sophisticated mechanisms of gene regulation to rapidly adapt its metabolism to changes in the external environment, to communicate with other cells in the ecological niche, or to escape the host immune response[20].

Urinary tract infection (UTI) is one of the most common infectious diseases in clinical practice. It is the second most common infection in humans, with its incidence only being lower than that of respiratory tract infections. The etiological agents most frequently involved in UTI are enterobacteria, non-fermenting bacteria, fungi, enterococci, and staphylococci. The most common and most important staphylococcal species related to UTI are *S. aureus* and *S. saprophyticus*; however, other CoNS species have gained importance in recent years. A study conducted in our laboratory by Ferreira et al. [44] identified *S. saprophyticus* as the most frequent species (56.4%) in the etiology of UTI, followed by *S. aureus* (16.9%), *S. epidermidis* (15.9%), *S. haemolyticus* (7.9%), *S. warneri* (1.9%), and *S. lugdunensis* (1.0%). *Staphylococcus aureus* is relatively uncommon as a causative agent of UTI in the general population [45], but can colonize and cause infection in some patients through the ascending route. Other factors such as urinary tract instrumentation and the presence of catheters increase the risk of UTI caused by this microorganism[46]. In our study [44], *S. aureus* was the second most common species among hospital isolates (23.5%) after *S. epidermidis* (41.1%). A high rate of *S. aureus* (16.2%) was also detected in non-hospital samples (obtained from outpatient clinics and health centers).

Staphylococcus saprophyticus is the second most common etiological agent of UTI in the community and is rarely isolated from hospitalized patients or as a contaminant of urine cultures [47]. Although in study of Ferreira et al. [44] three *S. saprophyticus* strains were isolated from hospitalized patients with UTI, including two patients in the obstetric ward and one in the urology ward of HC-FMB, the infections were not considered to be hospital-acquired after analysis of the records of these patients since they were not related to hospital procedures and the onset of clinical manifestations of infection occurred within less than 72 hours after admission.

Analysis of the age and gender of patients with UTIs caused by *S. saprophyticus* showed that 74 (73.2%) were women; of these, 56 (55.4%) were between 15 and 44 years of age. Braoios et al.[48] also observed a predominance of women among patients with UTI (69.1%) and most of them were between 20 and 49 years old (52.9%). Our data agree with the findings of these authors, demonstrating that the incidence of UTI caused by *S. saprophyticus* increases in adult life and that the infection mainly affects women, with a peak incidence at the beginning of sexual activity or related to it.

REFERENCES

[1] Kloos, WE; Bannerman, TL. Update on clinical significance of coagulase-negative Staphylococci. *Clin Microbiol Rev.*, 1994, 7(1), 117-40.

[2] Herbst, RH; Merricks, JW. *Staphylococcus albus* septicemia following nephrolithotomy: recovery with penicillin. *J Am Med Assoc.*, 1945, 127(9), 518-9.

[3] Harry; CH. *Staphylococcus albus haemolyticus* meningitis in an infant. *J Pediatr.*, 1946,104(5),557.

[4] Smith, IM; Beals, PD; Kingsbury, KR; Hasenclever, HF. Observations on *Staphylococcus albus* septicemia in mice and men. *AMA Arch Intern Med.*, 1958, 102(3), 375-88.

[5] Pulverer, G; Halswick, R. Coagulase-negative Staphylococci (*Staphylococcus albus*) as pathogens. *Dtsch Med Wochenschr.* 1967, 92(25), 1141-5.

[6] Pulverer; G. On the Pathogenicity of coagulase-negative Staphylococci. In: Jeljaszewicz J, editors. The Staphylococci: proceedings of V International Symposium on Staphylococci and Staphylococcal infections. Germany: Gustav Fischer Verlag Stuttgart,1985. 1-9.

[7] Wilson, TS; Stuart, RD. *Staphylococcus albus* in wound infection and in septicemia. *Can Med Assoc J.*, 1965, 93, 8-16.

[8] Pulverer, G; Pillich, J. Pathogenic significance of coagulase-negative *Staphylococci*. In: Finland M, Marget W, Bartmann K, editors. Bacterial infections: changes in their causative agents; trends and possible basis. New York: Springer-Verlag,1971. 91-96

[9] Pereira, AT. Coagulase-negative strains of *Staphylococcus* possessing antigen 51 as agents of urinary infection. *J Clin Pathol.*, 1962, 15:252-3

[10] Gallagher, DJ; Montogomerie, JZ; North, JD. Acute infections of the urinary tract and the urethral syndrome in general practice. *Br Med J.*, 1965, 1(5435), 622-6

[11] Mabeck, CE. Significance of coagulase-negative staphylococcal bacteriuria. *Lancet.* 1969, 2(7631), 1150-2.

[12] NNIS. National nosocomial infections surveillance system. Description of surveillance methods. *Am J infect Control.*, 1991, 19(1), 19-35.

[13] Wisplinghoff; H; Bischoff, T; Tallent, SM; Seifert, H; Wenzel, RP; Edmond, MB. Nosocomial bloodstream infections in US hospitals:

analysis of 24,179 cases from a prospective nationwide surveillance study. *Clin Infec Dis.*, 2004, 39(3), 309-14.

[14] Marra, AR; Camargo, LF; Pignatari, AC; Sukiennik, T; Behar, PR; Medeiros, EA; Ribeiro, J; Girão, E; Correa, L; Guerra, C; Brites, C; Pereira, CA; Carneiro, I; Reis, M; Souza, MA; Tranchesi, R; Barata, CU; Edmond, MB; Grupo de Estudo SCOPE Brasileiro. Nosocomial bloodstream infections in Brazilian hospitals: analysis of 2,563 cases from a prospective nationwide surveillance study. *J Clin Microbiol.*, 2011, 49, 1866-71.

[15] Hidron, AI; Edwards, JR; Patel, J; Horan, TC; Sievert, DM; Pollock, DA; Fridkin, SK; NHSN. Annual Update: Antimicrobial-Resistant Pathogens Associated with Healthcare-associated infections: Annual Summary of Data Reported to the National Healthcare Safety Network at the Centers for Disease Control and Prevention, 2006-2007. *Infect Control Hosp Epidemiol.*, 2008, 29(11), 996-1011.

[16] Assis, DB; Madalosso, G; Ferreira, SA; Yassuda, YY. Análise dos dados do Sistema de Vigilância de Infecção Hospitalar do Estado de São Paulo – Ano 2010. Centro de Vigilância Epidemiológica, Coordenadoria de Controle de Doenças, Secretaria de Estado da Saúde. Available at: http://www.cve.saude.sp.gov.br/htm/ih/pdf/ih11-dados10.pdf

[17] Sievert, DM; Ricks, P; Edwards, JR; Schneider, A; Patel, J; Srinivasan, A; Kallen, A; Limbago, B; Fridkin, S. Antimicrobial-Resistant Pathogens Associated with Healthcare-Associated Infections: Summary of Data Reported to the National Healthcare Safety Network at the Centers for Disease Control and Prevention, 2009–2010. *Infect Control Hosp Epidemiol.*, 2013, 34(1), 1-14.

[18] Grice, EA; Segre, JA. The skin microbiome. *Nat Rev Microbiol.*, 2011, 9(4), 244-53.

[19] Otto, M. *Staphylococcus epidermidis*-the 'accidental' pathogen. *Nature Reviews.*, 2009, 7(8), 556-7.

[20] Schoenfelder, SMK; Langea, S; Eckart, M; Hennig, S; Kozytska, S; Ziebuhr, W. Success through diversity – How *Staphylococcus epidermidis* establishes as a nosocomial pathogen. *Int J of Medical Microb.*, 2010, 300(6), 380–386.

[21] Massey, RC; Horsburgh, MJ; Lina, G; Hook, M; Recker, M. The evolution and maintenance of virulence in *Staphylococcus aureus*: a role for host-to-host transmission? *Nature Reviews.* 2009, 4(12), 953-8.

[22] Freeman, J; Platt, R; Epstein, MF; Smith, NE; Sidebottom, DG; Goldmann, DA. Birth weight and length of stay as determinants of nosocomial coagulase-negative staphylococcal bacteremia in neonatal intensive care unit populations: potential for confounding. *Am J Epidemiol.*, 1990, 132(6), 1130-40.

[23] Kacica, MA; Horgan, MJ; Ochoa, L; Sandler, R; Lepow, ML; Venezia, RA. Prevention of gram-positive sepsis in neonates weighing less than 1500 grams. *J Pediatr.*, 1994, 125(2), 253-8.

[24] Kirchhoff, LV; Sheagren, JN. Epidemiology and clinical significance of blood cultures positive for coagulase-negative *Staphylococcus*. *Infect Control.*, 1985, 6(12), 479-86.

[25] ANVISA. Agência Nacional de Vigilância Sanitária. Pediatria: prevenção e controle de infecção hospitalar. Brasília, Ministério da Saúde, 2006, 116.

[26] CDC. Centers for Disease Control and Prevention. Guidelines for the Prevention of Intravascular Catheter-Related Infections. 2011, 52(9), e162-93

[27] Sidebottom, DG; Freeman, J; Platt, R; Epstein, MF; Goldmann, DA. Fifteen-year experience with bloodstream isolates of coagulase-negative Staphylococci in neonatal intensive care. *J Clin Microbiol.*, 1988, 26(4), 713-8.

[28] Horan, TC; Andrus, M; Dudeck, MA. CDC/NHSN surveillance definition of health care-associated infection and criteria for specific types of infections in the acute care setting. *Am J Infect Control.*, 2008,36(5), 309-32.

[29] D'Angio, CT; McGowan, KL; Baumgart, S; St, Geme, J; Harris, MC. Surface colonization with coagulase-negative Staphylococci in premature neonates. *J Pediatr.*, 1989, 114(6), 1029-34.

[30] Shinefield, HR; St, Geme, III JW. Staphylococcal infections. In: Remington JS, Klein JO, editors. Infectious diseases of the fetus and newborn infant. Philadelphia: WB Saunders Co; 2001, 1217-47.

[31] Metzger-Maayan, A; Linder, N; Marom, D; Visher, T; Ashkenazi, S; Sirota, L. Clinical and laboratory impact of coagulase-negative Staphylococcal bacteremia in preterm infants. *Acta Paediatr.*, 2000,89(6), 690-3.

[32] Cunha, MLRS; Lopes, CAM; Rugolo, LMSS; Chalita, LVS. Significância clínica de estafilococos coagulase-negativa isolados de recém-nascidos. *J Pediatr (Rio J).*,2002,78(4),279-88.

[33] Silbert, S; Rosa, DD; Matte, U; Goldim, JR; Barcellos, SH; Procianoy, RS. Coagulase-negative *Staphylococcus* sp. in blood cultures from infants less than 60 days old: infection versus contamination. *J Pediatr (Rio J).*, 1997, 73(3), 161-5.

[34] Lowy, FD; Hammer, SM. *Staphylococcus epidermidis* infections. *Ann Intern Med.*, 1983, 99(6), 834-9.

[35] Isaacs, D. A ten year, multicentre study of coagulase negative staphylococcal infections in Australasian neonatal units. *Arch Dis Child Fetal Neonatal Ed.*, 2003, 88(2), F89–93.

[36] Vergnano, S; Menson, E; Kennea, N; Embleton, N; Russell, AB; Timothy, Watts, T; et al. Neonatal infections in England: the Neon IN surveillance network. *Arch Dis Child Fetal Neonatal Ed.*, 2011, 96(1), F9-F14.

[37] Davenport, A. Peritonitis remains the major clinical complication of peritoneal dialysis: The London, UK, Peritonitis Audit 2002–2003. *Perit Dial Int 2009*,29(3), 297-302.

[38] Perez Fontan, M; Rodriguez-Carmona, A; Garcia-Naveiro, R; Rosales, M; Villaverde, P; Valdes, F. Peritonitis-related mortality in patients undergoing chronic peritoneal dialysis. *Perit Dial Int.*, 2005, 25(3), 274-84.

[39] Barretti, P; Moraes, TMC; Camargo, CH; Caramori, JCT; Mondelli, AL; Montelli, AC; Cunha; MLRS.Peritoneal dialysis-related peritonitis due to *Staphylococcus aureus*: a single-center experience over 15 Years. *Plos One.*,2012,7, e31780.

[40] Troidle, L; Gorban-Brennan, N; Kliger, A; Finkeltein, F. Differing outcomes of gram-positive and gram-negative. *Am J Kidney Dis.*, 1998, 32(4), 623-8.

[41] Peacock, SJ; Howe, PA; Day, NP; Crook, DW; Winearls, CG; Berendt, AR. Outcome following staphylococcal peritonitis. *Perit Dial Int.*, 2000, 20(2), 215-9.

[42] Cunha, MLRS; Montelli, AC; Fioravante, AM; Batalha, JEN; Caramori, JCT; Barretti, P. Predictive factors of outcome following staphylococcal peritonitis in continuous ambulatory peritoneal dialysis. *Clin Nephrol.*, 2005,64(5),378-82.

[43] Barretti, P; Montelli, AC; Batalha, JEN; Caramori, JCT; Cunha, MLRS. The role of virulence factors in the outcome of staphylococcal peritonitis in CAPD patients. *BMC Infect Dis.*, 2009, 9, 212-9.

[44] Ferreira, AM; Bonesso, MF; Mondelli, AL; Cunha, MLRS. Identification of *Staphylococcus saprophyticus* isolated from

patients with urinary tract infection using a simple set of biochemical tests correlating with 16S 23S interspace region molecular weight patterns. *J Microbiol Methods.*, 2012, 91(3),406-11.

[45] Barrett, SP; Savage, MA; Rebec, MP; Guyot, A; Andrews, N; Shrimpton, SB. Antibiotic sensitivity of bacteria associated with community acquired urinary tract infection in Britain. *J Antimicrob Chemother.*, 1999, 44(3), 359-65.

[46] Coll, PP; Crabtree, BF; O'Connor, PJ; Klenzak, S. Clinical risk factors for methicillin-resistant *Staphylococcus aureus* bacteriuria in a skilled-care nursing home. *Arch Fam Med.*, 1994, 3(4), 357-60.

[47] Jordan, PA; Iravani, A; Richard, GA; Baer, H. Urinary tract infection caused by *Staphylococcus saprophyticus*. *J Infect Dis.*, 1980, 142(4), 510-15.

[48] Braoios, A; Turatti, TF; Meredija, LCS; Campos, TRS; Denadai, FHM. Infecções do trato urinário em pacientes não hospitalizados: etiologia e padrão de resistência aos antimicrobianos. *J Bras Patol Med Lab.*, 2009, 45(6), 449-6.

Chapter 3

CATHETER-RELATED INFECTIONS

The advent of intensive care centers has led to advances in the treatment of the critically ill patient, increasing survival even in high-risk populations such as patients with sepsis, immunodepressed patients, cancer patients, and very low weight preterm infants. However, the evolution of the therapeutic arsenal using increasingly more invasive techniques resulted in the breakdown of barriers and exposure of previously intact tissues, rendering them susceptible to infection. Catheter-related infections are an example of this scenario and occur when a germ invades the bloodstream through a vascular catheter. Infections associated with the use of intravascular devices correspond to 10 to 20% of all healthcare-associated infections and are a leading cause of morbidity and mortality. These infections are sources of bacteremia and sepsis in hospitalized patients and increase the length and costs of hospital stays and death rates. A Brazilian study using the SCOPE methodology identified the presence of a central venous catheter (CVC) as the main risk factor for the occurrence of nosocomial bloodstream infections in 70.3% of patients [1].

Approximately 65% of catheter-related infections result from the migration of microorganisms of the skin microbiota from the site of catheter insertion, 30% of cases of contamination are intraluminal, and 5% occur through other routes such as infusion of contaminated fluids and distant foci of infection [3].

More than 150 million vascular catheters are used annually in hospitals and clinics of the United States. Most of these devices are peripheral venous catheters; however, an expressive number of these catheters, more than 5 million, are CVCs inserted in deep or central vessels [4]. More than 200,000 bloodstream infections occur in the United States, most of them related to nontunneled CVCs. The purpose of these catheters is variable and includes the administration of medication, blood and blood derivatives, parenteral nutrition, monitoring of the patient's hemodynamic condition, and vascular access for hemodialysis [5].

In neonatal intensive care units (NICUs), the rate of infection is inversely proportional to the birth weight of the newborn and ranges from 9.1 per 1000 catheter days in infants with birth weights < 1,000 g to 3.5 per 1000 catheter days in infants with birth weights > 2,500g [69]. In Brazil, catheter-related bloodstream infections (CRBSIs) are the most common type of infection in NICUs. In the study of Pessoa-Silva et al. [6], the incidence of CRBSIs ranged from 17.3 per 1000 catheter days in infants weighing 1,501 to 2,500 g to 34.9 per 1000 catheter days in infants weighing < 1,000 g.

Most CRBSIs are caused by coagulase-negative staphylococci (CoNS). In the study of Perlman et al. [7], these microorganisms were responsible for 55.5% of CRBSIs in newborns, followed by *Staphylococcus aureus* (13.2%) and *Enterococcus* (9.2%). Gram-negative microorganisms such as the enterobacteria *Escherichia coli*, *Enterobacter* and *Klebsiella pneumoniae* and the non-fermenting Gram-negative bacilli *Pseudomonas aeruginosa* and *Acinetobacter baumannii* were isolated from 20.2% of these infections. *Staphylococcus epidermidis* was isolated from 39.8% of CRBSIs, *S. warneri* from 6.9% and other CoNS from 8.7%, with most CRBSIs being caused by Gram-positive microorganisms (79.8%). Data reported by the NHSN of the CDC [2], comprising the period from January 2006 to October 2007, also ranked CoNS first place in the etiology of CRBSIs (34.1%), followed by *Enterococcus* species (16%), *Candida* species (11.8%), and *S. aureus* in fourth place (9.9%). Gram-negative microorganisms including the enterobacteria *Escherichia coli*, *Enterobacter*, *Klebsiella pneumoniae* and *Klebsiella oxytoca* were responsible for 12.4% of these infections, and the non-fermenting Gram-negative bacilli *Pseudomonas aeruginosa* and *Acinetobacter baumannii* for 5.3%.

The pathogenesis of CRBSI is complex and only partially understood. It is speculated that these infections are directly related to adherence of the microorganisms, which then become able to colonize the device. There are four potential sources for catheter colonization and the occurrence of CRBSI: the insertion site of the catheter in the skin, the catheter hub, hematogenic contamination, and contamination of the infusion fluid. The skin is the main source of colonization and infection of short-term catheters. Bacteria present on the patient's skin migrate along the surface of the catheter and colonize the distal end, resulting in infection [8]. The hub is another source of colonization.

Microorganisms may be introduced through the hands of the healthcare team and then migrate along the inner surface of the catheter, causing bloodstream infection. Greater colonization of the inner surface can be expected in the case of prolonged use of the catheter hub (more than 30 days) [9]. Hematogenic contamination of CVCs from a distant focus of infection, such as pneumonia, gastrointestinal infection or urinary tract infection, has been suggested, but is not an important cause of catheter colonization and is rarely confirmed [10]. Parenteral nutrition solutions and lipid emulsions promote the growth of many bacteria and fungi, such as *Candida parapsilosis* and *Malassezia furfur* [11]. Although many outbreaks of nosocomial bacteremia were caused by contaminated infusion fluid, the contribution of this source to endemic primary nosocomial bacteremias is very low. All of these sources of contamination are important, but none of them competes with contamination by the patient's own microorganisms in areas near the catheter insertion site. This explains why CoNS are the microorganisms most frequently associated with these infections since they are the most common bacteria on the skin.

The adherence of microorganisms to the surface of catheters depends on the interaction of three factors: the host, the microorganism and the catheter material. The host reacts against the catheter, considering it a foreign body, and forms a sheath of fibrin and fibronectin around it. These host protein components that cover the surface of the catheter permit the adherence of *S. aureus*. These microorganisms produce coagulase which promotes thrombogenesis, in addition to various other proteins present on their surface, called microbial surface components recognizing adhesive matrix molecules (MSCRAMMs). The latter are adhesins that possess receptors for proteins released by the host, such as fibrinogen and fibronectin which permit adherence and colonization of the catheter [12, 13].

CoNS can colonize the native surface of the catheter, as well as surfaces conditioned by host proteins. Once these bacteria have adhered to the catheter surface, they proliferate, forming multiple layers, and produce an extracellular polysaccharide that forms a biofilm which potentiates their pathogenicity. The biofilm not only favors the adherence of microorganisms, but also their maintenance by acting as a barrier against the attack of antibiotics, neutrophils, phagocytes, macrophages, and antibodies. The concentration of an antibiotic required to kill bacteria in a biofilm is 100-1000 times greater than that necessary to eliminate the same species in suspension, a fact that makes treatment difficult and increases the possibility of recurrent infections since the bacteria are protected against the host immune system [14].

Candida is another microorganism of concern since studies have shown that this species is responsible for about 80% of fungal infections in the hospital environment. Furthermore, some species can produce a biofilm similar to the bacterial biofilm in the presence of glucose-containing solutions, a finding explaining the increase in the proportion of CRBSI caused by pathogenic fungi, especially among patients receiving parenteral nutrition [15].

The third important factor for microbial adherence is the catheter material. *In vitro* studies have shown that polyvinyl chloride or polyethylene catheters are less resistant to the adherence of microorganisms than catheters made of Teflon, silicone or polyurethane [16].

With respect to the length of hospital stay, death rate and hospital costs related to CRBSIs, all of these factors increase significantly. The occurrence of these infections prolongs hospitalization from 6.5 to 22 days and increases costs from US$29,000 to US$56,000 per episode of infection. It is estimated that mortality is 13% to 28% higher among patients under these conditions compared to patients of the same severity without this complication [17, 18].

Catheter-related sepsis in newborns also represents a serious complication since intravascular catheters are widely used for numerous procedures in the NICU, permitting rapid intravenous access for the administration of medications and parenteral nutrition, among others [19]. Although the use of catheters is a fundamental procedure for newborn survival in the NICU, it is also an important risk factor for the development of infection and can increase the incidence of bloodstream infections and, consequently, morbidity rates and the length of hospital stay [16].

As observed in adult individuals, about 60% of CRBSIs in newborns are caused by Gram-positive bacteria, particularly CoNS which are the main members of the skin and mucosal microbiota. These microorganisms have been recognized as the most frequent etiological agents of infections in newborns, especially among very low birth weight infants (< 1,500 g) [3].

There was no consensus among scholars regarding the diagnosis of catheter-related infections. In 2002, the CDC standardized these concepts, differentiating between significant catheter colonization and catheter-related sepsis. According to the CDC [16], significant catheter colonization is characterized by a strong correlation between local inflammation at the site of catheter insertion and the isolation of more than 15 colony-forming units (CFU) of a microorganism from a segment of the catheter. The term "local infection" is therefore frequently used in this situation. Other investigators use the term "colonization" to describe this event, in contrast to the term "contamination" when less than 15 CFU are isolated. However, a positive culture of a catheter segment in the absence of a positive blood culture should not be considered device-related bacteremia or fungemia.

The term sepsis or CRBSI is employed when, in addition to semiquantitative culture of a catheter segment showing growth ≥ 15 CFU or quantitative culture with growth ≥ 1,000 CFU, the same microorganism (species and antibiogram) is isolated from blood cultures and clinical signs of sepsis (fever, hypothermia, apnea) are present in the absence of an apparent source of infection, except for the catheter. Thus, the definitive diagnosis is established when the catheter is significantly colonized with the same microorganism found in the blood culture [16]. It should be noted that the lack of standardization of clinical and diagnostic criteria and the diverse concepts of infection compromise epidemiological surveillance systems of infection and impair generalization of the results of studies.

Within this context, before the development of the semiquantitative culture technique, most clinical microbiology laboratories used broth culture of catheter tips for qualitative assessment. This technique provided poorly reliable and nonspecific data that did not permit the distinction between colonization and infection. In a study conducted in our laboratory [20] comparing qualitative and semiquantitative culture for the diagnosis of CRBSI in newborns from the Neonatal Unit of the University Hospital of the Botucatu Medical School (Hospital das Clínicas da Faculdade de Medicina de Botucatu – HC-FMB), it was observed that, although less sensitive (90%), semiquantitative culture showed higher specificity (71%) than the qualitative technique (100% sensitivity and 60% specificity). The authors concluded that the catheter semiquantitative culture method has advantages for the diagnosis of CRBSI in newborns when compared to the traditional qualitative method.

The reliability of culture depends on the technique used. The semiquantitative culture method proposed by Maki et al. [21], in which a catheter segment is rolled over a blood agar plate, is the most common technique used to determine CRBSI rates and several studies have confirmed its importance. In their study, Maki et al. [21] demonstrated that semiquantitative culture of a catheter tip showing ≥ 15 CFU was better correlated with the presence of infection; however, the authors found only four cases of catheter-related sepsis, all of them related to the number of microorganisms with confluent growth. Analyzing these results, the authors discussed that the cut-off value of ≥ 15 CFU for a positive result required further evaluation to determine the number of CFU that shows the best correlation with the presence of CRBSI without reducing the sensitivity of the test.

The diagnosis of CRBSIs in neonatal units is done using methods that are similar to those employed for the diagnosis of these infections in adult patients according to the criterion proposed by Maki et al. [21]. Within this context, our group conducted a study to determine the cut-off of semiquantitative culture that best correlated with the presence of CRBSI in newborns [22]. Eighty-five catheter tips obtained from 63 newborns were studied. The microorganisms isolated from catheters and peripheral blood cultures were identified and submitted to antimicrobial susceptibility testing by the disk diffusion method. The optimal cut-off value was determined using a receiver-operating characteristic (ROC) curve. The gold standard was the definitive diagnosis of CRBSI based on the isolation of the same microorganism (species and antimicrobial susceptibility profile) from catheter and peripheral blood cultures in the absence of another apparent focus of infection, except for the catheter. Eight (72.7%) of the 11 episodes of infection diagnosed were associated with CoNS, six of them caused by *S. epidermidis*. On the ROC curve, the optimal cut-off for the diagnosis of catheter-related infection was ≥ 122 CFU. The sensitivity and specificity of this cut-off were 91.0% and 81.1%, respectively, versus 91.0% and 71.6% when a cut-off ≥ 15 CFU was used. The cut-off value of ≥ 122 CFU showed higher specificity and a higher positive predictive value (PPV), without loss of sensitivity, when compared to ≥ 15 CFU.

Studying adult patients, Collignon et al.[23] suggested ≥ 5 CFU as the best cut-off for the detection of CRBSI since this value showed higher sensitivity (92%) and the same specificity (83%) as those found for cultures with growth ≥ 15 CFU. In order to determine whether semiquantitative culture of catheter tips is useful for the diagnosis of CRBSI, a test with high specificity and high PPV is necessary and desirable; however, Collignon et al.[23] obtained a PPV of only 8.8%, with 124 catheters showing false-positive results. On the other hand, specificity would be 94% for a cut-off value ≥ 100 CFU, with a reduction in the number of false-positive results to 46. According to Brun-Buisson et al. [24], the authors should choose growth ≥ 100 CFU instead of ≥ 5 CFU as the best cut-off since a diagnostic test with a PPV less than 10% cannot be considered adequate for clinical use.

In our study, although most cases of CRBSI showed growth > 122 CFU, there was one case of infection caused by *S. aureus* whose culture showed growth of only 8 CFU. Other authors also reported CRBSI and semiquantitative cultures with growth < 15 CFU [23]. These results might be explained by the use of antibiotics before culture or by intraluminal contamination of the catheter. The latter is a limiting factor of semiquantitative culture which only detects microorganisms adhered to the outer surface of the device [25]. This technique is limited mainly in the case of long-term catheters in which the inner surface is the predominant source of colonization. This disadvantage does not apply to quantitative methods (vortex or sonication method). In the vortex method described by Brun-Buisson et al. [24], the catheter segment is flushed with sterile distilled water to detach microorganisms from the inner surface, followed by serial dilution and seeding on a blood agar plate. Growth of 1,000 CFU/ml or higher is indicative of CRBSI.

Considering the aspects described above and in view of the fact that semiquantitative culture only determines the presence of microorganisms on the outer surface of the catheter, whereas qualitative culture permits isolation of microorganisms from the outer surface and from the catheter lumen, another study is being conducted in our laboratory whose main objective is to compare the semiquantitative technique proposed by Maki et al. [21] and the quantitative method described by Brun-Bruisson et al. [24].

One disadvantage of these semiquantitative and quantitative culture methods is the need of catheter removal for culture. Thus, other diagnostic methods of CRBSI that do not require removal of the device are desirable. The most promising methods are those that permit maintenance of the access. This advantage is particularly striking in critically ill patients with difficult access and in patients with long-term catheters. One of these new techniques for the diagnosis of CRBSI without removal of the catheter is the differential time to positivity (DTP) method, which compares the time to positivity of qualitative cultures of blood collected from the CVC and a peripheral vein by automated continuous monitoring of the growth of microorganisms in the catheter and peripheral blood cultures. CRBSI is indicated if the catheter sample becomes positive 2 hours or more before the peripheral blood sample. Another technique is paired quantitative culture of blood collected from the catheter and peripheral vein. The growth of microorganisms that is at least five times greater in blood collected from the catheter than in blood obtained from a peripheral vein indicates a positive result [26].

DTP is a simpler method than quantitative blood culture and is widely available since many clinical microbiology laboratories have adopted the use of automated systems for blood culture monitoring. This technique could therefore be used in large prospective clinical trials, representing an easily adoptable low-cost method for the diagnosis of CRBSI which does not require removal of the catheter [27]. In the study of Blot et al. [27], the diagnosis of

CRBSI was obtained in 16 of 17 patients with a positive culture result of blood collected from the CVC at least 2 hours before the peripheral blood culture, with 91% sensitivity and 94% specificity.

For the elucidation of CRBSI, it is fundamental to determine the similarity of strains of microorganisms isolated from catheter tips and blood cultures. Biochemical tests can determine the gender and species and the antibiogram permits to evaluate similarities between strains isolated from the catheter and blood culture based on the resistance profile of the bacteria present in the samples, contributing not only to the treatment of CRBSIs but also to their diagnosis (Figure 1). In a comparative study of molecular techniques and microbiological methods for the identification of sources of nosocomial infections, Martín-Lozano et al. [28] evaluated the usefulness of antibiograms for epidemiological studies and concluded that conventional clinical and/or microbiological criteria for the diagnosis of the source of bacteremia, including antibiograms, are not always precise and sufficient. Several of the problems encountered are the result of the variable expression of the phenotypic characteristics used as parameters.

For these reasons, significant efforts have been made to develop alternative methods that combine ease of use, reliability and low cost. Thus, molecular typing techniques are now available that can provide additional discriminatory power, mainly because they do not depend on the expression of genes for evaluation. Some of these methods are based on the principles of the polymerase chain reaction (PCR), such as random amplified polymorphic DNA-PCR (RAPD-PCR) and repetitive extragenic palindromic sequence-based PCR (REP-PCR). RAPD-PCR is basically a variation of the PCR protocol with two distinct characteristics: it uses a single primer instead of a primer pair and the single primer has an arbitrary sequence, i.e., its target sequence is unknown [29]. REP-PCR uses primers complementary to naturally occurring, highly conserved, repetitive and noncoding sequences (generally 30 to 500 bp) [30]. The advantages of these methods include their low cost and relative simplicity when compared to other techniques.

In an attempt to develop more specific and sensitive methods that are able to establish genetic relationships between isolates obtained from catheters and blood cultures, a study conducted by Pazzini [31] in our laboratory analyzed the genomic profile of microorganisms isolated from catheters and blood cultures using PCR-based techniques and compared the isolated strains based on clustering with similarity coefficients for the diagnosis of CRBSI in newborns.

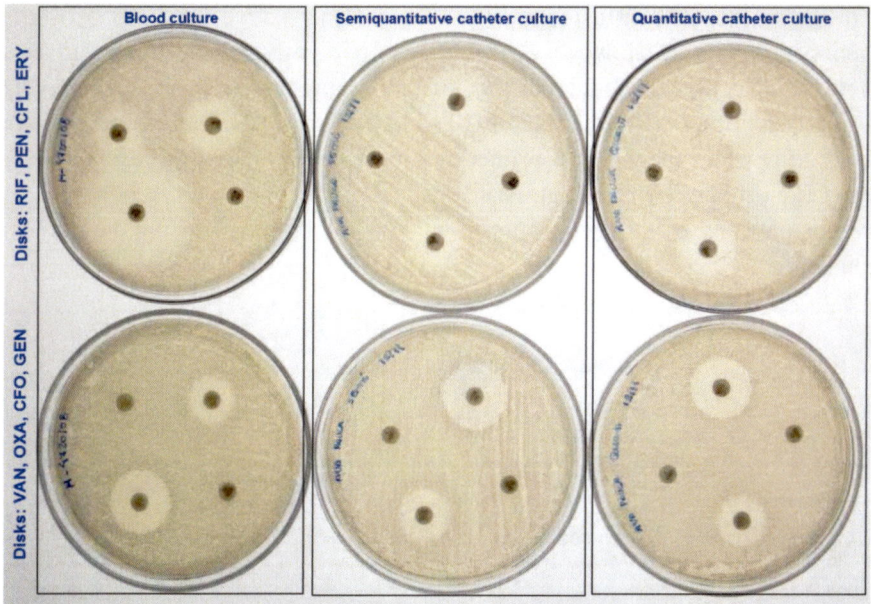

Figure 1. Antimicrobial susceptibility testing: agar disk diffusion method for the determination of similarity between coagulase-negative staphylococci isolated from catheters and blood cultures. VAN: vancomycin; OXA: oxacillin; CFO: cefoxitin; GEN: gentamicin; RIF: rifampicin; PEN: penicillin; CFL: cephalothin; ERY: erythromycin.

Random primers from Operon Technology, Inc. (OPT13, OPR18, OPR13, OPK18 and OPERON21) and Invitrogen (RAPD1, RAPD7, M13, 1026, Random primer, ERIC1 and ERIC2) were tested to standardize the RAPD reactions. The most reproducible primers with the greatest differentiation capacity that produced well-defined and strong bands for each group of microorganisms were selected for the study (Figure2). Reference strains of the international American Type Culture Collection (ATCC) and strains of the same species studied, but unrelated (e.g., strains obtained from another hospital), were used to evaluate the differentiation capacity of the primers. Primers presenting at least six well-defined bands were selected. For better reproducibility of RAPD-PCR, two primers were chosen for each genus or species of microorganism studied.

All 21 isolates that were positive for CRBSI in the phenotypic test (isolates from the catheter and blood culture presenting the same susceptibility profile by the disk diffusion method) were also positive for **CRBSI** by the genotypic methods. However, 10 isolates were positive in the genotypic tests and negative in the phenotypic test; of these, seven were positive by RAPD using two primers (RAPD 1 and M13) and by REP-PCR, three were positive by RAPD using the RAPD1 primer, and only one isolate was positive using the M13 primer. There was no statistically significant difference between the results obtained by the disk diffusion method and by RAPD using the primers studied.

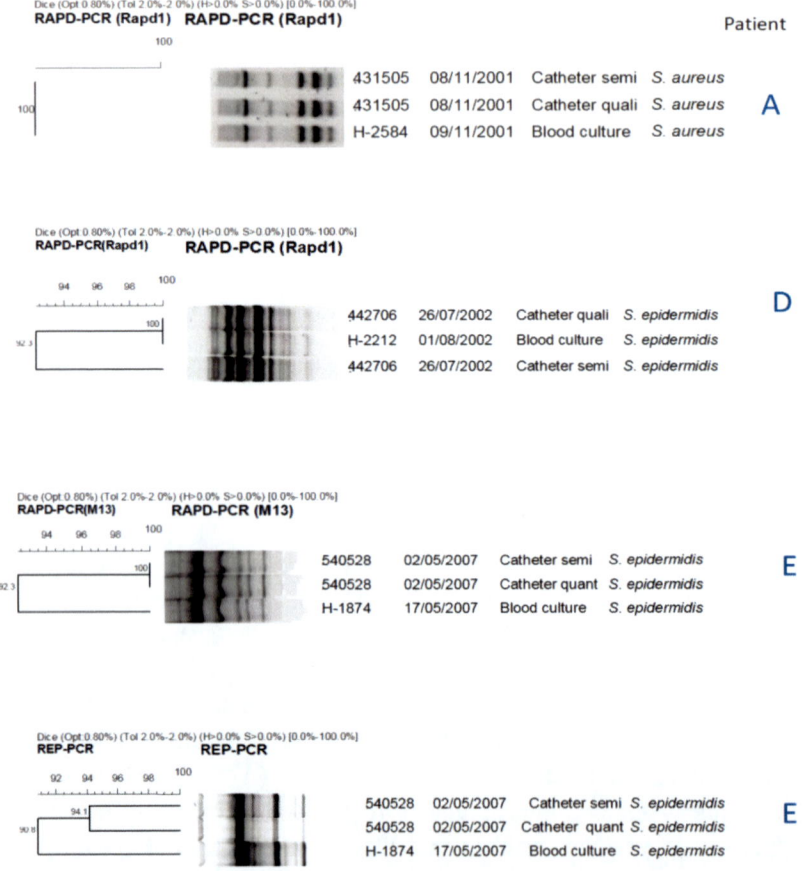

Figure 2. Dendrograms obtained by random amplified polymorphic DNA-PCR (RAPD-PCR) and repetitive extragenic palindromic sequence-based PCR (REP-PCR) for *Staphylococcus* spp. isolated from catheters and blood cultures. *S. aureus* with the primer RAPD1 (patient A), with a similarity coefficient of 100%. *S. epidermidis* with the primer RAPD 1 (patient D), by RAPD-PCR using primer M13 (patient E), and by REP-PCR (patient E), with a similarity coefficient of 90 to 99.9%.

The results showed significant detection of bacteria of the genus *Staphylococcus* by all methods used for the evaluation of CRBSI ($p < 0.0001$). With respect to *Staphylococcus* species, there was a higher frequency of *S. epidermidis* associated with these infections ($p < 0.0001$).

According to Tenover et al. [32], the combination of two or more typing methods increases the discriminatory potential of the patterns of the strains studied. In this respect, despite the differences observed between techniques, RAPD-PCR and REP-PCR were important for the confirmation of CRBSI. Furthermore, the use of two RAPD-PCR reactions employing different primers attenuated the problem of the low reproducibility of the method.

The determination of the clonal profile of the *S. epidermidis* strains isolated from blood cultures by RAPD-PCR and REP-PCR demonstrated the persistence of *S. epidermidis* clones for 1 to 7 years in the NICU of HC-FMB. All major clusters contained isolates associated with CRBSIs. In some situations, minor clusters also contained isolates associated with these infections. Huebner et al. [33], Neumeister et al. [34] and Villari et al. [35] also detected persistent clones of *S. epidermidis* in the hospital environment. Villari et al. [35] suggested that a significant portion of infections caused by *S. epidermidis* can be attributed to transmission among patients and that certain strains can become endemic over long periods of time.

These results confirm the importance of cross-infection with CoNS in the NICU. The success of predominant clones in these studies may be related to still unknown factors that provide advantages to these microorganisms in terms of colonization or their capacity to infect patients. These factors probably include the formation of a biofilm and resistance to antimicrobial drugs.

In a study comparing molecular methods for the typing of *S. aureus* isolates, Tenover et al. [32] reported some important parameters of good molecular techniques, such as reproducibility, discriminatory power, ease of use, and easy interpretation of the techniques, in addition to stronger band intensity which helps with the interpretation of the results [36]. In our study, RAPD-PCR and REP-PCR showed good discriminatory power since they differentiated the strains studied from ATCC strains and from unrelated isolates, with a similarity coefficient ≤ 70. However, RAPD-PCR using the RAPD1 and M13 primers for *Staphylococcus* spp. provided better band patterns than REP-PCR, facilitating interpretation of the results. Determination of the clonal profile of *S. epidermidis* strains isolated from blood cultures revealed greater clustering of isolates associated with CRBSIs, in addition to easier application of the RAPD-PCR method when compared to REP-PCR.

Despite reports of low reproducibility of RAPD-PCR [37], no difficulty was observed in our study.

The molecular typing techniques showed an additional discriminatory power, especially in the case of infections caused by CoNS. This finding is of great importance since these microorganisms are part of the normal microbiota and the understanding of the relationships between these organisms is fundamental for the elucidation of CRBSIs.

REFERENCES

[1] Marra, AR; Camargo, LF; Pignatari, AC; Sukiennik, T; Behar, PR; Medeiros, EA; Ribeiro, J; Girão, E; Correa, L; Guerra, C; Brites, C; Pereira, CA; Carneiro, I; Reis, M; Souza, MA; Tranchesi, R; Barata, CU; Edmond, MB; GrupodeEstudoSCOPEBrasileiro. Nosocomial bloodstream infections in Brazilian hospitals; analysis of 2,563 cases from a prospective nationwide surveillance study. *J Clin Microbiol.*, 2011, 49, 1866-71.

[2] Hidron, AI; Edwards, JR; Patel, J; Horan, TC; Sievert, DM; Pollock, DA; Fridkin, SK. NHSN. Annual Update; Antimicrobial-Resistant Pathogens Associated with Healthcare-associated infections; Annual Summary of Data Reported to the National Healthcare Safety Network at the Centers for Disease Control and Prevention, 2006-2007. *Infect Control Hosp Epidemiol.*, 2008, 29(11), 996-1011.

[3] NNIS. National Nosocomial Infections Surveillance. System report, data summary from January 1992 to June 2004, issued October 2004. *Am J Infect Control.*, 2004, 32, 470-85.

[4] Percival, SL; Kite, P; Eastwood, K; Murga, R; Carr, J; Arduino, MJ; et al. Tetrasodium EDTA as a novel central venous catheter lock solution against biofilm. *Infect Control Hosp Epidemiol.*, 2005, 26(6), 515-9.

[5] Bacuzzi, A; Cecchin, A; Del, Bosco, A; Cantone, G; Cuffari, S. Recommendations and reports about central venous catheter-related infection. *Surg Infect* (Larchmt)., 2006, 7 Suppl 2, S65-7.

[6] Pessoa-Silva, CL; Richtmann, R; Calil, R; Santos, R; Costa, MLM; Frota, ACC; et al. Healthcare-associated infections among neonates in neonatal units in Brazil. *Infect Control Hosp Epidemiol.*, 2004, 25, 772-777.

[7] Perlman, SE; Saiman, L; Larson, EL. Risk factors for late-onset health care-associated bloodstream infections in patients in neonatal intensive care units. *Am J Infect Control.*, 2007, 35 (3), 177–82.

[8] Maki, DG. Infection caused by intravascular devices; Pathogenesis, strategies for prevention. London, England; Royal Society of Medicine Services Limited, 1991.

[9] Raad, I; Costerton, W; Sabharwal, U; Sacilowski, M; Anaissie, E; Bodey, GP. Ultrastructural analysis of indwelling vascular catheters; a quantitative relationship between luminal colonization and duration of placement. *J Infect Dis.*, 1993, 168(2), 400-7.

[10] Kovacevich, DS; Faubion, WC; Bender, JM; Schaberg, DR; Wesley, JR. Association of parenteral nutrition catheter sepsis with urinary tract infections. *J Parenter Enteral Nutr.*, 1986, 10(6), 639-41.

[11] Plouffe, JF; Brown, DG; Silva, J; Eck, T; Stricof, RL; Fekety, FR. Nosocomial outbreak of *Candida parapsilosis* fungemia related to intravenous infusions. *Arch Intern Med.*, 1977, 137(12), 1686-9.

[12] Vaudaux, P; Pittet, D; Haeberli, A; Lerch, PG; Morgenthaler, JJ; Proctor, RA; et al. Fibronectin is more active than fibrin or fibrinogen in promoting *Staphylococcus aureus* adherence to inserted intravascular catheters. *J Infect Dis.*, 1993, 167(3), 633-41.

[13] Arciola, CR; Baldassarri, L; Montanaro, L. Presence of *icaA* and *icaD* genes and slime production in a collection of staphylococcal strains from catheter-associated infections. *J Clin Microbiol.*, 2001, 39(6), 2151-6.

[14] Stewart, PS; Costerton, JW. Antibiotic resistance of bacteria in biofilms. *Lancet.*, 2001, 358(9276), 135-8.

[15] Andes, D; Nett, J; Oschel, P; Albrecht, R; Marchillo, K; Pitula, A. Development and characterization of an in vivo central venous catheter *Candida albicans* biofilm model. *Infect Immun.*, 2004, 72(10), 6023-31.

[16] CDC. *Centers for Disease Control and Prevention.* O'Grady, NP; Alexander, M; Dellinger, EP; Gerberding, JL; Heard, SO; Maki, DG; Masur, H; McCormick, RD; Mermel, LA; Pearson, ML; Raad, II; Randolph, AWeinstein. Healthcare infection control practices advisory committee. *Guidelines for the Prevention of Intravascular Catheter-Related Infections. Infect Control Hosp Epidemiol*,2002, 23(12), 759-69.

[17] Collignon, PJ; Wilkinson, IJ; Gilbert, GL; Grayson, ML; Whitby, RM. Health care-associated *Staphylococcus aureus* bloodstream infections; a

clinical quality indicator for all hospitals. *Med J Aust.*, 2006, 184(8), 404-6.

[18] Warren, DK; Yokoe, DS; Climo, MW; Herwaldt, LA; Noskin, GA; Zuccotti, G; et al. Preventing catheter-associated bloodstream infections; a survey of policies for insertion and care of central venous catheters from hospitals in the prevention epicenter program. *Infect Control Hosp Epidemiol.*, 2006, 27(1), 8-13.

[19] Schmidt-Sommerfeld, E; Snyder, G; Rossi, TM; Lebenthal, E. Catheter-related complications in 35 children and adolescents with gastrointestinal disease on home parenteral nutrition. *JPEN J Parenter Enteral Nutr.*, 1990, 14(2), 148-51.

[20] Marconi, C; Cunha, MLRS; Lyra, JC; Bentlin, MR; Batalha, JEN; Sugizaki, MF; Rugolo, LMSS. Comparison between qualitative and semiquantitative catheter-tip cultures; laboratory diagnosis of catheter-related infection in newborns. *Braz J Microbiol.*, 2008, 39(2), 262-7.

[21] Maki, DG; Weise, CE; Sarafin, HW. A semiquantitative culture method for identifying intravenous-catheter-related infection. *N Engl J Med.*, 1977, 296(23), 1305-9.

[22] Marconi, C; Cunha, MLRS; Lyra, JC; Bentlin, MR; Batalha, JEN;, Sugizaki, MF; Corrente, JE; Rugolo, LMSS. Usefulness of catheter tip culture in the diagnosis of neonatal infections. *J Pediatr (Rio J.)*, 2009, 85(1), 80-3.

[23] Collignon, PJ; Soni, N; Pearson, IY; Woods, WP; Munro, R; Sorrell, TC. Is semiquantitative culture of central vein catheter tips useful in the diagnosis of catheter-associated bacteremia? *J Clin Microbiol.*, 1986, 24(4), 532-5.

[24] Brun-Buisson, C; Abrouk, F; Legrand, P; Huet, Y; Larabi, S; Rapin, M. Diagnosis of central venous catheter-related sepsis. Critical level of quantitative tip cultures. Arch Intern Med., 1987, 147(5), 873-7.

[25] Bouza, E; San, Juan, R; Muñoz, P; Pascau, J; Voss, A; Desco, M; et al. A European perspective on intravascular catheter-related infections; report on the microbiology workload, aetiology and antimicrobial susceptibility (ESGNI-005 Study). *Clin Microbiol Infect.*, 2004, 10(9), 838-42.

[26] Capdevila, JA; Planes, AM; Palomar, M; Gasser, I; Almirante, B; Pahissa, A; et al. Value of differential quantitative blood cultures in the diagnosis of catheter-related sepsis. *EurJ Clin Microbiol InfectDis.*, 1992, 11(5), 403-7.

[27] Blot, F; Nitenberg, G; Chachaty, E; Raynard, B; Germann, N; Antoun, S; et al. Diagnosis of catheter-related bacteraemia; a prospective comparison of the time to positivity of hub-blood versus peripheral-blood cultures. *Lancet.*, 1999, 354(9184), 1071-7.
[28] Martín-Lozano, D; Cisneros, JM; Becerril, B; Cuberos, L; Prados, T; Ortíz-Leyba, C; et al. Comparison of a repetitive extragenic palindromic sequence-based PCR method and clinical and microbiological methods for determining strain sources in cases of nosocomial *Acinetobacter baumannii* bacteremia. *J Clin Microbiol.*, 2002, 40(12), 4571-5.
[29] Williams, JG; Kubelik, AR; Livak, KJ; Rafalski, JA; Tingey, SV. DNA polymorphisms amplified by arbitrary primers are useful as genetic markers. *Nucleic Acids Res.*, 1990, 18(22), 6531-5.
[30] Healy, M; Huong, J; Bittner, T; Lising, M; Frye, S; Raza, S; et al. Microbial DNA typing by automated repetitive-sequence-based PCR. *J Clin Microbiol.*, 2005, 43(1), 199-207.
[31] Pazzini, LT. Caracterização genotípica de microrganismos isolados de infecções da corrente sanguínea relacionadas a cateteres em recém-nascidos [Dissertation]. Botucatu; Universidade Estadual Paulista Júlio de Mesquita Filho,2010.
[32] Tenover, FC; Arbeit, R; Archer, G; Biddle, J; Byrne, S; Goering, R; et al. Comparison of traditional and molecular methods of typing isolates of *Staphylococcus aureus*. *J Clin Microbiol.*, 1994, 32(2), 407-15.
[33] Huebner, J; Pier, GB; Maslow, JN; Muller, E; Shiro, H; Parent, M; et al. Endemic nosocomial transmission of *Staphylococcus epidermidis* bacteremia isolates in a neonatal intensive care unit over 10 years. *J Infect Dis.*, 1994, 169(3), 526-31.
[34] Neumeister, B; Kastner, S; Conrad, S; Klotz, G; Bartmann, P. Characterization of coagulase-negative staphylococci causing nosocomial infections in preterm infants. *EurJ Clin Microbiol InfectDis.*, 1995, 14(10), 856-63.
[35] Villari, P; Sarnataro, C; Iacuzio, L. Molecular epidemiology of *Staphylococcus epidermidis* in a neonatal intensive care unit over a three-year period. *J Clin Microbiol.*, 2000, 38(5), 1740-6.
[36] Kremer, K; vanSoolingen, D; Frothingham, R; Haas, WH; Hermans, PW; Martín, C; et al. Comparison of methods based on different molecular epidemiological markers for typing of *Mycobacterium tuberculosis* complex strains; interlaboratory study of discriminatory power and reproducibility. *J Clin Microbiol.*, 1999, 37(8), 2607-18.

[37] Penner, GA; Bush, A; Wise, R; Kim, W; Domier, L; Kasha, K; et al. Reproducibility of random amplified polymorphic DNA (RAPD) analysis among laboratories. *PCR Methods Appl.*, 1993, 2(4), 341-5.

Chapter 4

IDENTIFICATION OF *STAPHYLOCOCCUS* SPP.

Despite the increase in the clinical significance of coagulase-negative staphylococci (CoNS), in most routine laboratories staphylococci are identified exclusively based on colony morphology, Gram staining and the production of catalase and coagulase. These characteristics only permit the classification of staphylococci isolated from clinical samples into *S. aureus* and non-*S. aureus*, the latter being simply classified as CoNS. However, in view of the increased occurrence of infections caused by different CoNS species, it becomes increasingly important to learn more about the epidemiology, antimicrobial resistance and pathogenic potential of individual species. This is particularly relevant for strains isolated from blood cultures since it is often difficult to determine the clinical significance of a CoNS isolate.

Kloos and Schleifer [1] designed a scheme that can be used to easily differentiate CoNS species based on their biochemical characteristics. The scheme proposed by these authors and modified by Bannerman[2] is the conventionally used method. However, this method is relatively time consuming for use in routine clinical microbiology laboratories because of the large number of biochemical tests.

The precise identification of CoNS is important to predict the pathogenic potential of each species and its antibiotic susceptibility profile, thus permitting more adequate evaluation of the clinical significance of each species. There are divergences in the specific literature regarding the clinical importance of CoNS identification. According to some authors [3], this procedure is not clinically significant, whereas others [4] believe that this identification is important for the differentiation between contamination and infection. In this respect, the identification of CoNS is important to associate certain species with specific infections [5], since data suggest that, in addition to *S. epidermidis* and *S. saprophyticus* which are considered to be pathogenic, some species such as *S. haemolyticus*, *S. lugdunensis* and *S. schleiferi* are more frequently associated with infections than other species [6, 7]. Repeat isolates of CoNS from patients with invasive diseases should be identified to permit the comparison of strains. Furthermore, the identification of species is a prerequisite before starting procedures for epidemiological studies.

The development of methods for the identification of staphylococcal species and subspecies permits to obtain information about the variety of CoNS present in clinical samples and about their role as etiological agents of infectious processes. In recent years, several commercial systems for the rapid identification of staphylococci have been developed as an alternative to classical identification protocols [2]. However, these diagnostic systems present problems such as high cost and a long incubation time and often provide unreliable results [8, 9]. Additionally, many of these kits were designed to identify all known CoNS species (isolated from clinical, veterinary and food samples) and are therefore not very specific. On the basis of these considerations and the need for rapid, simple and reliable methods, a study was conducted to compare four techniques for the identification of CoNS, i.e., a reference method [1, 2], the commercial mini-API Staph system and two methods modified in our laboratory. In addition, a simplified scheme was proposed to develop alternative identification methods that combine reliability, simplicity and low cost, especially for facilities with limited resources [10].

Two identification methods modified in our laboratory were used (simplified method and disk method). The simplified method was divided into two steps. In the first step, the fermentation of xylose, sucrose, trehalose, maltose and mannitol, production of hemolysin, and anaerobic growth in thioglycolate were tested. The tests used in the second step varied according to the results obtained in the first step after 72 hours of incubation at 37°C. The 100 staphylococcal isolates obtained from clinical samples were tested simultaneously by the four methods proposed. International reference strains (ATCC) of *S. epidermidis*, *S. simulans*, *S. saprophyticus* and *S. xylosus* were identified correctly by the four methods. Among the *S. warneri* reference strains, only strain ATCC 10209 was misidentified by the API Staph system as *S. saprophyticus* (%ID = 58.7%), as *S. hominis* (%ID = 19.5%), or as *S. warneri* (%ID = 15.8%). The simplified method performed in two steps did not differ from the reference method in terms of the identification of CoNS species. Inaccurate identification by the API Staph system was observed for *S.*

epidermidis (2.2%), *S. warneri* (47.1%), *S. hominis* (25%), and *S. haemolyticus* (37.5%). The sensitivity and specificity of the simplified method were 100% for all species studied. The disk method showed a sensitivity of 93.8% in the identification of *S. hominis* due to the lack of fermentation of sucrose on the disk, which led to the classification of this strain as *S. caprae*, and 100% sensitivity for the other species, whereas specificity was 98.8% for *S. hominis*, 98.9% for *S. caprae*, and 100% for the other species. The simplified method using the identification scheme proposed resulted in the identification of *S. epidermidis*, *S. hominis*, *S. xylosus*, *S. capitis* and *S. simulans* in a single step using a total of seven biochemical tests. This number is smaller than the number of tests employed in the reference method (16 tests). Since *S. epidermidis* is the most frequently isolated species, 70 to 90% of strains isolated in the clinical laboratory can be identified using a reduced number of tests.

The commercial API Staph kit showed the lowest accuracy in the identification of CoNS among the methods studied (84% agreement), as also reported by Bannerman et al. [11] and Renneberg et al. [12]. *Staphylococcus warneri* and *S. hominis* were the species most difficult to identify. Bannerman et al. [11] also reported lower accuracy in the identification of these species. In the study of Ieven et al. [13], *S. hominis* was identified with the least accuracy by the API Staph ID 32 system. This finding can be explained by the lack of complementary tests such as resistance to novobiocin, anaerobic growth in thioglycolate and hemolysin production. Three (3%) of the 100 strains analyzed by the API Staph system were erroneously identified as *S. aureus*, a fact also reported by Renneberg et al. [12]. The kit was found to be inefficient in these cases since it does not require the result of the fundamental and most widely accepted test for the identification of *S. aureus*, i.e., the coagulase test [14].

In most routine laboratories, *S. saprophyticus* is identified mainly based on resistance to novobiocin (5 µg), absence of hemolysis, and a negative coagulase and/or DNAse test. However, it has been recognized that other CoNS species, including *S. cohnii*, *S. sciuri*, *S. xylosus* and *S. hominis* subsp. *novobiosepticus*, are also resistant to novobiocin at this concentration[14, 15]. These results suggest that additional tests, including the fermentation of carbohydrates, should be performed in conjunction with the novobiocin susceptibility test for the correct identification of CoNS species [16].

In recent years, several commercial systems for staphylococcal identification have been developed as an alternative to the classical identification protocol of Bannerman [2], which is very laborious and time consuming for use in routine laboratory practice. Automated systems and commercial kits based on miniature biochemical tests are nowadays widely used both in routine laboratories and in research. However, these diagnostic systems present problems in terms of cost and incubation time and, more importantly, are still unable to reliably distinguish different CoNS species due to the variable expression of phenotypic characteristics. Additionally, many of the automated systems and kits are based on colorimetric results and their subjective interpretation can lead to ambiguity [17].

Bannerman et al. [11] evaluated the updated database of the Vitek I Gram-Positive Identification (GPI) card (Biomérieux) using 500 clinical isolates. Overall agreement between the GPI card and conventional methods was 89%. The card identified 92% of the *S. epidermidis* isolates studied, 95% of the *S. haemolyticus* isolates, 88% of the *S. capitis* subsp. *capitis* isolates, and 100% of the *S. saprophyticus* isolates. Microorganisms not included in the database, such as *S. lugdunensis*, were either misidentified or not identified by the GPI card. In the study of Perl et al. [9], the GPI card correctly identified only 67% of 185 isolates. The authors emphasized that the poor performance of the GPI card was probably due to the predominance of non-*S. epidermidis* species among the 227 isolates tested.

Another technique used for the identification of staphylococci is the polymerase chain reaction (PCR). This technique permits the genotypic identification of different staphylococcal species with high sensitivity and specificity. ITS-PCR permits the analysis of intergenic transcribed spacer (ITS) regions between 16S and 23S ribosomal RNA (rRNA) gene loci, a technique frequently used in PCR fingerprinting for the identification and discrimination of bacterial strains at the species and subspecies level [18]. On polyacrylamide or agarose gel, the amplified regions form a band pattern that is specific for each species. The strains investigated are then compared to the respective patterns of ATCC strains, leaving no doubt regarding the accurate identification of different staphylococcal species. This method has been described originally by Barry et al.[19] and Jensen et al.[20], who also studied the identification of *Staphylococcus* and successfully applied the technique to the differentiation of strains of four staphylococcal species: *S. aureus*, *S. epidermidis*, *S. saprophyticus*, and *S. warneri*.

Over the last 10 years, ITS-PCR has been widely used for the typing of bacterial strains because of the marked polymorphism in ITS regions [20, 21]. This polymorphism is the result of the presence of transfer RNA, which is responsible for the length and sequence of the regions between operons and is similar in the same species [20, 22]. The study of the polymorphism rate in

ITS regions (16S and 23S rRNA) was found to be useful both epidemiologically and taxonomically, amplifying the 16S and 23S rRNA regions using primers G1 and L1 described by Jensen et al.[20].

Couto et al.[17] applied ITS-PCR to identify 600 staphylococcal strains originating from different hospitals using 29 reference strains (ATCC) of recognized *Staphylococcus* species as positive control. The 29 staphylococcal species showed a unique ITS-PCR pattern, with the method proving to be rapid and safe for the identification of staphylococci in clinical samples, providing highly reliable and reproducible results.

DNA analyses have been the methods of choice for the identification of microorganisms because of their greater specificity and sensitivity. CoNS strains not identified at the species level or misidentified by conventional phenotypic tests can be correctly identified by genotypic methods. Despite a reduction in the costs of molecular techniques in recent years, these methods continue to be expensive for routine use in clinical microbiology laboratories. Therefore, significant efforts have been made to develop alternative identification methods, combining speed, reliability and low cost. A study conducted in our laboratory [23] compared three phenotypic identification methods for staphylococci isolated from urinary infections, including the simplified scheme of biochemical tests proposed in a previous study, novobiocin disk susceptibility testing, and the automated Vitek I system. Genotypic identification by ITS-PCR was used as the reference method. Among 101 strains studied, all 17 *S. aureus* strains were positive for coagulase, DNAse and the *coa* gene. However, two of the 84 CoNS isolates were DNAse positive, but were negative for coagulase and the *coa* gene. The remaining 82 strains were negative for coagulase, DNAse and the *coa* gene. The DNAse test showed 100% sensitivity and 97.6% specificity and the coagulase test showed the same sensitivity (100%) and specificity (100%) as the reference genotypic method (*coa* gene). Agreement was 98% for the DNAse test and 100% for the coagulase test.

Comparison of the identification methods for all species studied showed sensitivity and specificity of 89.1% for the novobiocin disk, of 81.2% and 92%, respectively, for the Vitek I system, and of 98% for the simplified method of biochemical tests. Regarding the identification of *S. saprophyticus*, the sensitivity and specificity of the simplified method of biochemical tests and of the novobiocin susceptibility test were 100%, whereas the sensitivity and specificity of the Vitek I system were 84.2% and 100%, respectively. Despite the rapid identification (2 to 15 hours) and the large number of biochemical tests (29 tests), the Vitek I system failed to identify the CoNS species most commonly found in urine samples (*S. saprophyticus*), mainly because of the lack of detection of novobiocin resistance in some isolates (6/9), as well as *S. epidermidis*, the second most frequent CoNS species in urine and the most common in other clinical materials due to failure in the sucrose test (5/5). With respect to the two species erroneously identified by the simplified method of biochemical tests, *S. haemolyticus* showed an incorrect result in the urease test. The same isolate also gave a positive urease reaction on the Vitek I card, suggesting that rare strains of *S. haemolyticus* can be positive for urease. *Staphylococcus epidermidis* presented a positive trehalose test, a finding that may have been due to possible contamination with the sugar since the same isolate gave a negative trehalose reaction on the Vitek I card.

Kim et al. [24] compared the results of 120 clinical CoNS strains identified by the Vitek 2 and Microseq 500 systems (Applied Biosystems). The latter is a commercially available system for analysis of the 16S rRNA gene. The Vitek 2 system correctly identified 105 (87.5%) of the isolates and misidentified six (5.0%). When the low-level discrimination results that included the correct identification were considered together, the agreement rate was 95.0% (114/120). Therefore, even when a current device and updated software were used, the agreement rates between the genotypic identification method and the automated method were lower than those between the simplified method of biochemical tests and the genotypic method (98.0%) found in our study. These findings confirm that, although faster, the automated systems are still unable to reliably distinguish between different CoNS species.

REFERENCES

[1] Kloos, WE; Schleifer, KH. Simplified scheme for routine identification of human *Staphylococcus* species. *J Clin Microbiol.*, 1975, 1(1), 82-8.

[2] Bannerman, TL. *Staphylococcus, Micrococcus,* and other catalase-positive cocci that grow aerobically. In: Murray PR, Baron EJ, Jorgensen JH, Pfaller MA, Yolken RH, editors. Manual of Clinical Microbiology. Washington: American Society Microbiology,2003, 384-404.

[3] Oren, I; Merzbach, D. Clinical and epidemiological significance of species identification of coagulase-negative Staphylococci in a microbiological laboratory. *Isr J Med Sci.*, 1990, 26(3), 125-8.

[4] Lowy, FD; Hammer, SM. *Staphylococcus epidermidis* infections. *Ann Intern Med.*, 1983, 99(6), 834-9.

[5] Rupp, ME; Archer, GL. Coagulase-negative Staphylococci: pathogens associated with medical progress. *Clin Infect Dis.*, 1994, 19(2), 231-43.

[6] Herchline, TE; Ayers, LW. Occurrence of *Staphylococcus lugdunensis* in consecutive clinical cultures and relationship of isolation to infection. *J Clin Microbiol.*, 1991, 29(3), 419-21

[7] Low, DE; Schmidt, BK; Kirpalani, HM; Moodie, R; Kreiswirth, B; Matlow, A; et al. An endemic strain of *Staphylococcus haemolyticus* colonizing and causing bacteremia in neonatal intensive care unit patients. *Pediatrics.*, 1992, 89(4 Pt 2), 696-700.

[8] Grant, CE; Sewell, DL; Pfaller, M; Bumgardner, RV; Williams, JA. Evaluation of two commercial systems for identification of coagulase-negative Staphylococci to species level. *Diagn Microbiol Infect Dis.*, 1994, 18(1), 1-5.

[9] Perl, TM; Rhomberg, PR; Bale, MJ; Fuchs, PC; Jones, RN; Koontz, FP; et al. Comparison of identification systems for *Staphylococcus epidermidis* and other coagulase-negative *Staphylococcus* species. *Diagn Microbiol Infect Dis.*, 1994, 18(3), 151-5.

[10] Cunha, MLRS; Sinzato, YK; , Silveira, LVA. Comparison of methods for the identification of coagulase-negative staphylococci. *Mem Inst Oswaldo Cruz.*, 2004,99(8),855-60.

[11] Bannerman, TL; Kleeman, KT; Kloos, WE. Evaluation of the Vitek Systems Gram-Positive Identification card for species identification of coagulase-negative Staphylococci. *J Clin Microbiol.*, 1993, 31(5), 1322-5.

[12] Renneberg, J; Rieneck, K; Gutschik, E. Evaluation of Staph ID 32 system and Staph-Zym system for identification of coagulase-negative Staphylococci. *J Clin Microbiol.*, 1995, 33(5), 1150-3.

[13] Ieven, M; Verhoeven, J; Pattyn, SR; Goossens, H. Rapid and economical method for species identification of clinically significant coagulase-negative Staphylococci. *J Clin Microbiol.*, 1995, 33(5), 1060-3.

[14] Koneman, EW; Allen, SD; Janda, WM; Schreckenberger, PC. Color Atlas and Textbook of Diagnostic Microbiology. 5th ed. Philadelphia: J.B. Lippincott,1997.

[15] Hussain, Z; Stoakes, L; Stevens, DL; Schieven, BC; Lannigan, R; Jones, C. Comparison of the MicroScan system with the API Staph-Ident system for species identification of coagulase-negative Staphylococci. *J Clin Microbiol.*, 1986, 23(1), 126-8.

[16] Cunha, MLRS; Lopes, CML. Estudo da produção de beta -lactamase e sensibilidade às drogas em linhagens de estafilococos coagulase-negativos isolados de recém-nascidos. *J Bras Patol Med Lab.*, 2002, 38(4), 281-290.

[17] Couto, I; Pereira, S; Miragaia, M; Sanches, IS; de Lencastre, H. Identification of clinical staphylococcal isolates from humans by internal transcribed spacer PCR. *J Clin Microbiol.*, 2001, 39(9), 3099-103.

[18] Daffonchio, D; Cherif, A; Brusetti, L; Rizzi, A; Mora, D; Boudabous, A; et al. Nature of polymorphisms in 16S-23S rRNA gene intergenic transcribed spacer fingerprinting of *Bacillus* and related genera. *Appl Environ Microbiol.*, 2003, 69(9), 5128-37.

[19] Barry, T; Colleran, G; Glennon, M; Dunican, LK; Gannon F. The 16s/23s ribosomal spacer region as a target for DNA probes to identify eubacteria. *PCR Methods Appl.*, 1991, 1(1), 51-6.

[20] Jensen, MA; Webster, JA; Straus, N. Rapid identification of bacteria on the basis of polymerase chain reaction-amplified ribosomal DNA spacer polymorphisms. *Appl Environ Microbiol.*, 1993, 59(4), 945-52.

[21] Gürtler, V; Stanisich, VA. New approaches to typing and identification of bacteria using the 16S-23S rDNA spacer region. *Microbiology.*, 1996, 142 (Pt 1), 3-16.

[22] Gürtler, V. The role of recombination and mutation in 16S-23S rDNA spacer rearrangements. *Gene.*, 1999, 238(1), 241-52.

[23] Ferreira, AM; Bonesso, MF; Mondelli, AL; Cunha MLRS. Identification of *Staphylococcus saprophyticus* isolated from patients with urinary tract infection using a simple set of biochemical tests correlating with 16S 23S interspace region molecular weight patterns. *J Microbiol Methods.*, 2012,91(3),406-11.

[24] Kim, M; Heo, SR; Choi, SH; Kwon, H; Park, JS; Seong, MW; et al. Comparison of the MicroScan, VITEK 2, and Crystal GP with 16S rRNA sequencing and MicroSeq 500 v2.0 analysis for coagulase-negative Staphylococci. *BMC Microbiol.*, 2008, 8, 233.

Chapter 5

STAPHYLOCOCCAL BIOFILMS

The virulence factors produced by coagulase-negative staphylococci (CoNS) and how these factors contribute to the pathogenicity of foreign body-associated infections are under investigation. Evidence indicates that pathogenicity is related to the production of an extracellular polysaccharide that permits the adhesion of these microorganisms to smooth surfaces. These microorganisms thus colonize catheters, heart valves, pacemakers and prosthetic joints, forming a biofilm [1]. The latter protects the bacteria against host immune defense mechanisms and antimicrobial drugs, either directly by preventing these agents from penetrating the bacterial cell or indirectly by maintaining the cell in an inactive resting state. For these reasons, biofilm formation is considered the main virulence factor of CoNS and the most important infections caused by these microorganisms are those involving foreign bodies [2].

Many authors define biofilms as associations of microorganisms and their extracellular products that adhere to biotic or abiotic surfaces. Biofilms are complex microbial ecosystems formed by populations that originate from a single or from multiple species. In most natural environments, the biofilm consists of multiple species. In contrast, in biomaterial-associated infections approximately 80% of the cells are *S. epidermidis*. This phenomenon can be explained by the easy access of this skin inhabitant to catheters and implants [3].

The adhesion of *S. epidermidis* depends on the physicochemical properties of the polymeric and bacterial surfaces. Since plastic surfaces are hydrophobic and the main determinant of adhesion is cell surface hydrophobicity, primary adhesion of *S. epidermidis* is similar for many of the biomaterials investigated [4].

When a certain material is implanted into an individual, body fluids such as serum proteins and platelets start to cover the implant, modifying its surface properties and facilitating bacterial adhesion[5, 6]. *Staphylococcus epidermidis* and *S. aureus* express dozens of proteins on their surface, called microbial surface components recognizing adhesive matrix molecules (MSCRAMMs), which specifically bind to extracellular matrix proteins of the host such as fibrinogen, collagen, fibronectin, and vitronectin. The most widely studied MSCRAMM is Fbe, also known as SdrG, a protein that facilitates the binding of these microorganisms to fibrinogen [7]. Depending on the growth conditions, biofilm-producing strains have also been shown to contain considerable amounts of extracellular teichoic acids [8], which increase the adhesion to surfaces lined with fibronectin. A probable role of these biofilm components in the virulence of *S. epidermidis* has been suggested [9].

The second stage of biofilm formation is characterized by the multiplication of CoNS in a monolayer of cells attached to the plastic material or host. For the formation of bacterial layers, the cells will bind to one another through polysaccharide intercellular adhesin (PIA), also called poly-N-acetylglucosamine, a polysaccharide that favors the adhesion between cells [10, 11]. PIA is a linear homopolymer of up to 130 β-1,6-N-acetylglucosamine residues found on the cell surface, which is composed of two polysaccharide fractions: polysaccharide I (> 80%) which contains 15 to 20% of deacetylated residues and is therefore positively charged, and polysaccharide II (< 20%) which is structurally related to polysaccharide I, but has a low content of non-N-acetylated glucosamine residues and contains phosphate and ester-linked succinate, rendering it anionic [3, 11, 12].

The production of PIA (Figure 3) is mediated by the products of the chromosomal *ica*(*intercellular adhesion*) gene which are organized in an operon structure. This operon contains the *icaADBC* genes (biosynthesis), in addition to the *icaR* gene which exerts a regulatory function and is transcribed in the opposite direction. Once this operon is activated, four proteins that are necessary for the synthesis of PIA are transcribed, i.e., IcaA, IcaD, IcaB and IcaC [11]. PIA is synthesized from UDP-N-acetylglucosamine by N-acetylglucosamine transferase, an enzyme encoded by the intercellular adhesion locus (*ica*), particularly *icaA*. The single expression of this gene induces low enzymatic activity, producing small amounts of the polysaccharide. However, simultaneous expression of *icaA* and *icaD* promotes a significant increase in N-acetylglucosamine transferase, which catalyzes the formation of oligomers of approximately 10-20 residues of β-1,6-N-acetylglucosamine [11, 13].

Another possibility proposed is that the protein encoded by *icaA*, a transmembrane protein consisting of approximately 412 amino acids [14], requires *icaD* to assume an active conformation. Mack et al. [14] demonstrated that the product of the *icaC* gene, a hydrophobic integral membrane protein consisting of approximately 355 amino acids, is necessary for the synthesis of N-acetylglucosamine oligomers to react with PIA antibodies. The *icaC* gene, when expressed concomitantly with *icaA* and *icaD*, induces the synthesis of longer oligomers with 130 residues and is probably also involved in the export of the nascent PIA chain. After export, PIA is deacetylated by protein IcaB which introduces positive charges that are crucial for its superficial localization and biological function. The production and deacetylation of PIA have been recognized as key factors in the virulence of *S. epidermidis* [15, 16] and are the main mechanisms involved in biofilm accumulation [11].

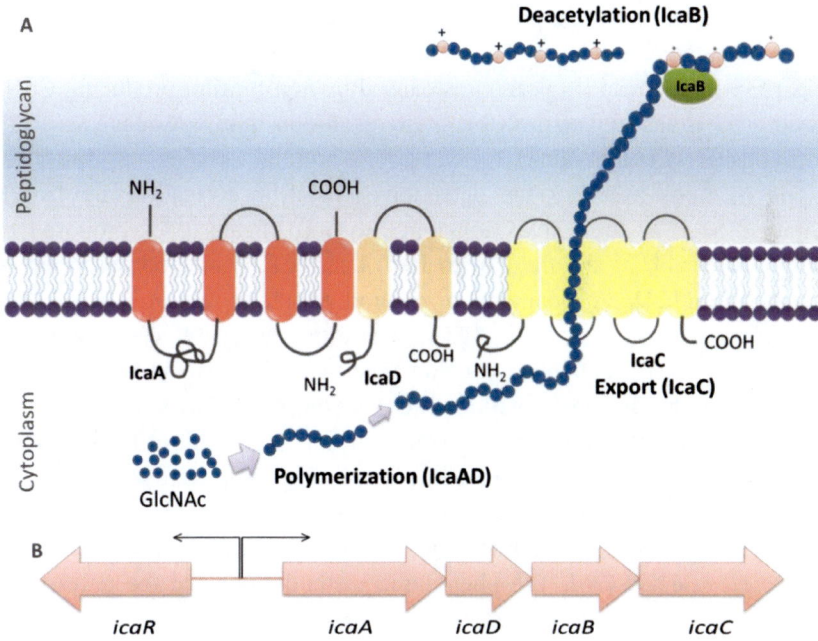

Figure 3. **A** Model for the biosynthesis of polysaccharide intercellular adhesin (PIA); **B** *ica* operon.

Several authors have emphasized biofilm formation as an epidemiological marker of infection [1, 17, 18]. On the other hand, other investigators found no association between biofilm-producing strains and the occurrence of infections caused by these microorganisms [19-21]. On the basis of the above considerations, we decided to evaluate the virulence of CoNS strains isolated from newborns hospitalized in the Neonatal Unit of the University Hospital of the Botucatu Medical School (Hospital das Clínicas da Faculdade de Medicina de Botucatu - HC-FMB), Brazil [22]. The study showed that a small proportion of isolates which produced biofilms (22.2%) was associated with infective processes; however, a significantly higher proportion of biofilm-producing *S. epidermidis* strains was isolated from blood cultures and foreign body than from secretions. Biofilm-producing CoNS strains have been isolated more frequently from patients with sepsis than from patients without invasive disease, indicating the risk for invasive infections in patients carrying biofilm-producing CoNS [18].

In another study designed to evaluate risk factors associated with the lack of peritonitis resolution in patients undergoing continuous ambulatory peritoneal dialysis, biofilm production was found to be an independent risk factor associated with the lack of resolution of peritonitis cases caused by CoNS [23]. In this respect, peritonitis caused by non-biofilm-producing CoNS showed a 27 times higher probability of resolution than episodes caused by biofilm-positive CoNS strains.

The methods used for the detection of biofilm formation are largely qualitative, such as the test of adherence to borosilicate tubes proposed by Christensen et al.[17] and the Congo Red Agar (CRA) method described by Freeman et al.[24]. A quantitative method is the polystyrene plate test described by Christensen et al. [25]. In addition to these tests, molecular methods such as the polymerase chain reaction (PCR) are used to amplify genes involved in biofilm formation. A collection of 80 CoNS strains isolated from clinical materials of newborns seen in the Neonatal Unit of HC-FMB and 20 strains isolated from the nares of healthy individuals were studied in our laboratory regarding biofilm production. Three phenotypic methods for biofilm detection in CoNS and detection of the *icaA*, *icaD* and *icaC* genes by PCR were evaluated [26]. The phenotypic methods included the polystyrene plate test, borosilicate tube assay, and the CRA method.

Using the borosilicate tube test (Figure 4), 82 of the 100 strains tested were found to be biofilm producers; of these, 44 strains were isolated from catheter tips, 23 from blood cultures, and 15 from nares. Analysis of each species showed that 70 (85.4%) *S. epidermidis* strains were positive. Biofilm formation was also observed in *S. warneri* (n = 4), *S. cohnii* (n = 3), *S. xylosus* (n = 2), *S. saprophyticus* (n = 2), and *S. lugdunensis* (n = 1). The borosilicate tube test showed 100% sensitivity and 100% specificity when compared to PCR (concomitant presence of the *icaA* and *D* genes or *icaACD*).

Investigation of biofilm production on polystyrene plates using a 540-nm filter (Figure 5) showed 96% sensitivity and 94% specificity when compared to PCR (concomitant presence of the *icaA* and *D* genes or *icaACD*).

According to the CRA test proposed by Freeman et al. [24], biofilm producers form black colonies on CRA, whereas non-producers develop red colonies. Seventy-six of the 100 CoNS strains studied formed black colonies on CRA (Figure 6); of these, 44 formed shiny black colonies and 32 dry black colonies. The colony color of the remaining 24 isolates ranged from pink (2%) and red (5%) to bordeaux (17%). It was therefore necessary to adopt a scale of five colors and the results of the CRA plates were compared to PCR in order to correlate the variation in colony color with the presence of the *ica* genes. Colonies showing a shiny black and dry black color were thus classified as positive strains and those showing a red, pink or bordeaux color as negative. Color scales were also adopted in other studies using CRA to improve the diagnosis of biofilm production; however, these studies used color variations that differed from the scale proposed in the study of our group. Vogel et al. [1] and Arciola et al. [27] classified strains appearing as very black, black and almost black colonies as positive and those appearing as bordeaux, red and very red colonies as negative. The CRA method showed 89% sensitivity and 100% specificity when compared to the concomitant presence of the *icaA* and *icaD* genes or *icaACD*.

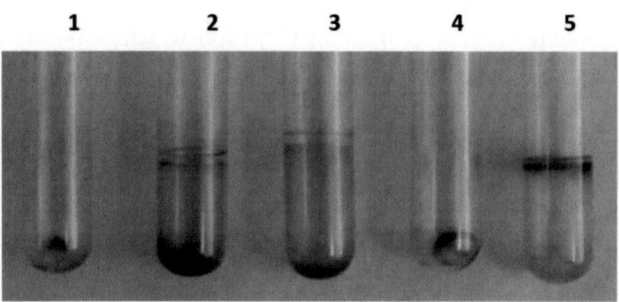

Figure 4. Investigation of biofilm production by coagulase-negative staphylococci in borosilicate test tubes. Lane 1: non-biofilm producer; lanes 2 and 3: biofilm producers; lane 4: *S. epidermidis* ATCC 12228 (negative control); lane 5: *S. xylosus* ATCC 29979 (positive control).

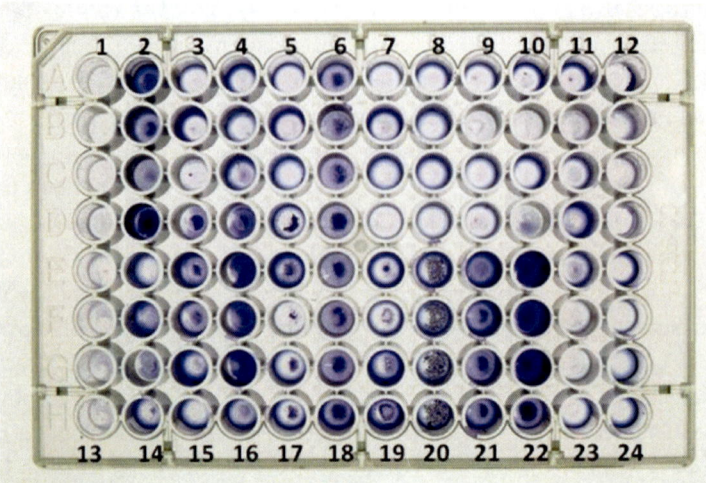

Figure 5. Investigation of biofilm production by coagulase-negative staphylococci on polystyrene plates. Column 1, rows A, B, C, and D (sterile TSB); column 13, rows E, F, G, and H: *S. epidermidis* ATCC 12228 (negative control); column 2, rows A, B, C, and D: *S. xylosus* ATCC 35984 (positive control); columns 3, 7, 8, 10, 11, and 17: strains classified as weakly adherent; columns 4, 6, 14, 15, 16, 18, 19, 20, 21, and 22:strains classified as strongly adherent; columns 5, 9, 12, 23, and 24: non-adherent strains.

The results of our study revealed no difference in the frequency of biofilm production between strains isolated from clinical samples and from the nostrils of healthy individuals for all tests used for detection of biofilm formation. Other authors also reported similar biofilm formation by CoNS strains isolated from different sources, including clinical specimens, the environment and the microbiota of healthy individuals [28, 29].

The results revealed one strain that tested positive by the polystyrene plate test and negative for the presence of the *ica* genes, a finding indicating PIA-independent biofilm formation. In the case of PIA-independent biofilm formation, adhesive proteins may replace the polysaccharide. The most important protein involved in this type of biofilm formation is accumulation-associated protein (Aap) [30].

In the study of Rohde et al. [31], 27% of biofilm-producing strains isolated from orthopedic device infections formed PIA-independent biofilms. In most cases, biofilm formation appeared to be mediated by Aap. Aap is a 220-kDa protein that needs to be proteolytically cleaved to a smaller form of 140 kDa to induce biofilm formation [32].

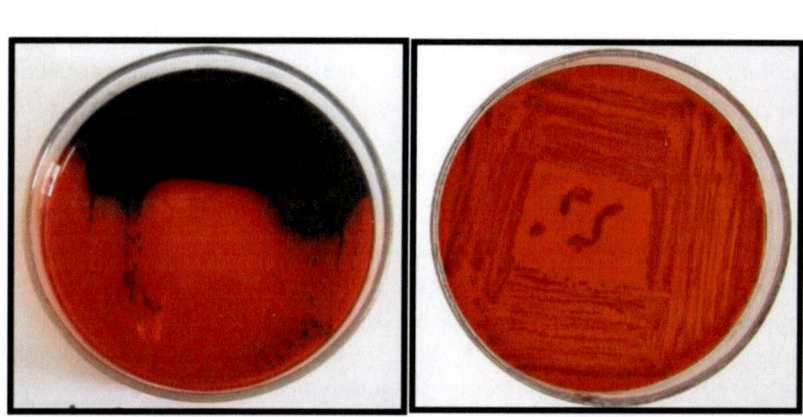

Figure 6. Investigation of biofilm production by coagulase-negative staphylococci on Congo red agar (CRA). **A** *S. simulans* ATCC 27851 (positive control) forming shiny black colonies on CRA. **B** *S. cohnii* (negative control) forming red colonies on CRA.

It has also been suggested that the expression of biofilm-associated protein (Bap) induces biofilm formation even in the absence of the *icaADBC* operon, highlighting the importance of this protein for the establishment of biofilms. There is evidence indicating a significant role of Bap in mastitis caused by *S. aureus* in cattle [33]. A *bap* homologous gene, called *bhp*, occurs in *S. epidermidis* in a similar manner as *bap* in strains isolated from animals. However, the mechanism whereby Bap and, eventually Bhp, contribute to biofilm formation is still unclear [33, 34].

Recently, the significance of extracellular DNA (eDNA) for the formation of biofilms has been demonstrated in *S. epidermidis* [35]. The importance of eDNA as a structural biofilm component has been shown for the first time in *Pseudomonas aeruginosa* [36], but was subsequently reported for a variety of bacterial species. Several studies suggest that the release of eDNA by *S. epidermidis* is mainly mediated by the activity of autolysin Atle (*atle* gene) [37]. Extracellular DNA is required for initial surface adhesion and subsequent maturation of the biofilm. Treatment of *S. epidermidis* cells with DNAse I inhibited the early stage of biofilm formation, suggesting that the release of eDNA contributes to the primary attachment of *S. epidermidis* to surfaces. In *S. aureus*, cell lysis and the release of eDNA have been associated with the presence of the *cidA* gene, a murein hydrolase regulator, whereas the production of thermonuclease has been shown to play a role in eDNA degradation and biofilm dispersion [38].

The biofilm guarantees the establishment of a communication system that coordinates metabolic activities for mutual benefit, as well as the production of virulence factors that facilitate the dissemination of these microorganisms in the host. One factor that contributes to the survival of bacteria inside the host, permitting efficient control of their virulence mechanisms including biofilm formation, is the so-called quorum sensing. Quorum sensing is a process of cell-cell communication in which signaling molecules, called autoinducers, regulate the behavior of bacteria according to population density. This process permits bacteria to perceive a time of high cell density to express virulence factors and to guarantee that the attack against the host is efficient. When a small number of bacteria release an autoinducer into the environment, its concentration is too low to be detected; however, at high cell density the autoinducer concentrations reach a level that is sufficient for cells to respond to the stimuli, activating or repressing target genes. This system therefore permits bacteria to coordinate their behavior based on environmental conditions, adapting themselves according to food availability, temperature, pH and osmolarity, as well as to regulate the expression of virulence factors, antibiotic production and biofilm formation, among others [11].

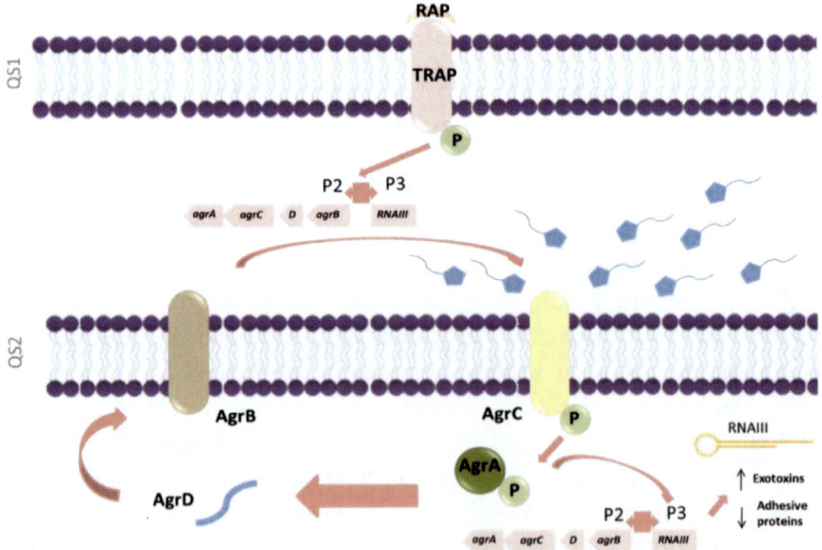

Figure 7. Interaction between the two quorum-sensing (QS) systems: QS1 - TRAP system and QS2 – *agr* system.

Staphylococcus aureus regulates its virulence factors through two quorum sensing (QS) systems, which regulate each other [39, 40]. QS1 consists of the activation of RNAIII through RNAIII-activating protein (RAP), a 21-kDa protein that induces the phosphorylation of the target protein of RAP (TRAP) [40, 41]. As bacteria multiply, RAP reaches a concentration threshold (during the mid-exponential growth phase) and induces the phosphorylation of its target molecule, TRAP [42]. The phosphorylation of TRAP activates the synthesis of products of the accessory gene regulator (*agr*) system (QS2). At the *agr* locus, two promoters in opposite orientations, P2 and P3, produce two transcripts (RNAII and RNAIII, respectively). RNAIII is a 510-nucleotide RNA molecule responsible for the transcription of genes of many virulence factors, such as extracellular toxins, enzymes and cell surface proteins present in *S. aureus* [43]. RNAII encodes four genes, *agrA*, *agrB*, *agrC* and *agrD*, arranged in an operon (*agr*), which together induce the synthesis of RNAIII. The products of the *agrB* and *agrD* genes (AgrB and AgrD, respectively) combine to produce an autoinducing peptide (AIP). AgrC is a transmembrane receptor for AIP. When AIP binds to AgrC, the latter phosphorylates and activates protein AgrA. Phosphorylated AgrA activates the P2 and P3 promoters. The synthesis of RNAIII is induced when the concentration of AIP, which is specific in the medium, reaches certain levels which are generally detected during passage from the exponential to the stationary growth phase. The final product of the *agr* regulatory cascade is RNAIII, an mRNA that functions as an inducer or repressor of accessory genes [40, 41, 43].

The QS1 and QS2 systems interact with one another (Figure 7) since the interplay between the phosphorylation of TRAP and AgrC by their respective autoinducers, RAP or AIP, regulates the expression of adhesion molecules or toxins, activating the expression of exotoxins at high cell density and the suppression of adhesins, with consequent dissemination of the microorganisms through the host tissue. TRAP has been shown to be a key molecule in the regulation of pathogenesis, since when its expression is inhibited by mutagenesis or inhibitory peptides, bacteria do not form biofilms, do not produce toxins and do not cause diseases [44, 45].

A peptide (YSPWTNF-NH$_2$), known at RNAIII-inhibiting peptide (RIP), can inhibit the activity of RNAIII, preventing the phosphorylation of TRAP and interfering with the quorum sensing system. RIP has been found to be effective in the treatment of medical device-associated staphylococcal infections, including those caused by multidrug-resistant *S. aureus* and *S. epidermidis* strains [42, 46]. A study using a rat graft model to evaluate biofilm production and the effect of RIP showed that injection of RIP into rats inhibited RNAIII, preventing the phosphorylation of TRAP and interfering with the quorum sensing system. As a result, biofilm production and infection rates were low [47]. Other studies demonstrated that, in addition to be highly effective in the prevention of device-associated staphylococcal infections by interfering with the quorum sensing system, RIP also reduces bacterial burden and can be useful for the treatment of infected wounds, representing an interesting future alternative to conventional antibiotics [48].

With respect to antibiotics, bacteria in biofilms generally exhibit increased minimum inhibitory concentrations, resulting in reduced susceptibility. Several factors contribute to this increased antimicrobial resistance; for example, the biofilm mode of growth permits bacteria to survive exposure to different antimicrobial drugs at up to 1,000-fold greater concentrations than those that would kill planktonic bacteria [49]. The relative resistance to antimicrobial drugs is probably due to the limited penetration of these agents since some parts of the biofilm are difficult to reach or metabolic activity is slow. The activity of some antimicrobial drugs is reduced in oxygen-deprived environments, a fact that also contributes to resistance of the biofilm since the availability of oxygen is lower at deeper levels [50]. Another factor could be the expression of certain genes inside the biofilm that confer antimicrobial resistance among Gram-positive species, especially staphylococci. This is a matter of concern when considering the use of antibiotics and highlights the need for new agents [49, 51, 52].

The fact that CoNS are a leading cause of medical device-related infections, as well as the adhesion capacity of these microorganisms and consequent biofilm formation, rendering the bacteria resistant to multiple antimicrobial drugs, makes the treatment of biofilm infections a difficult and expensive endeavor.

REFERENCES

[1] Vogel, L; Sloos, JH; Spaargaren, J; Suiker, I; Dijkshoorn, L. Biofilm production by *Staphylococcus epidermidis* isolates associated with catheter related bacteremia. *Diagn Microbiol Infect Dis.*, 2000, 36(2), 139-141.

[2] Cafiso, V; Bertuccio, T; Santagati, M; Campanile, F; Amicosante, G; Perilli, MG; et al. Presence of the *ica* operon in clinical isolates of

Staphylococcus epidermidis and its role in biofilm production. *Clin Microbiol Infect.*, 2004, 10(12), 1081-8.

[3] Götz, F. *Staphylococcus* and biofilms. *Mol Microbiol.*, 2002, 43(6), 1367-78.

[4] Vacheethasanee, K; Temenoff, JS; Higashi, JM; Gary, A; Anderson, JM; Bayston, R; et al. Bacterial surface properties of clinically isolated *Staphylococcus epidermidis* strains determine adhesion on polyethylene. *J Biomed Mater Res.*, 1998, 42(3), 425-32.

[5] Patti, JM; Allen, BL; McGavin, MJ; Höök, M. MSCRAMM-mediated adherence of microorganisms to host tissues. *Annu Rev Microbiol.*, 1994, 48, 585-617.

[6] vonEiff, C; Peters, G; Heilmann, C. Pathogenesis of infections due to coagulase-negative Staphylococci. *Lancet Infect Dis.*, 2002, 2(11), 677-85.

[7] Bowden, MG; Chen, W; Singvall, J; Xu, Y; Peacock, SJ; Valtulina, V; et al. Identification and preliminary characterization of cell-wall-anchored proteins of *Staphylococcus epidermidis*. *Microbiology.*, 2005, 151(Pt 5), 1453-64.

[8] Sadovskaya, I; Vinogradov, E; Flahaut, S; Kogan, G; Jabbouri, S. Extracellular carbohydrate-containing polymers of a model biofilm-producing strain, *Staphylococcus epidermidis* RP62A. *Infect Immun.*, 2005, 73(5), 3007-17.

[9] Hussain, M; Heilmann, C; Peters, G; Herrmann, M. Teichoic acid enhances adhesion of *Staphylococcus epidermidis* to immobilized fibronectin. *Microb Pathog.*, 2001, 31(6), 261-70.

[10] Mack, D; Fischer, W; Krokotsch, A; Leopold, K; Hartmann, R; Egge, H, et al. The intercellular adhesin involved in biofilm accumulation of *Staphylococcus epidermidis* is a linear beta-1,6-linked glucosaminoglycan: purification and structural analysis. *J Bacteriol.*, 1996, 178(1), 175-83.

[11] Otto, M. Staphylococcal biofilms. *Curr Top Microbiol Immunol.*, 2008, 322, 207-28.

[12] Morales, M; Méndez-Alvarez, S; Martín-López, JV; Marrero, C; Freytes, CO. Biofilm: the microbial "bunker" for intravascular catheter-related infection. *Support Care Cancer.*, 2004, 12(10), 701-7.

[13] McCann MT, Gilmore BF, Gorman SP. *Staphylococcus epidermidis* device-related infections: pathogenesis and clinical management. *J Pharm Pharmacol.*, 2008, 60(12), 1551-71.

[14] Mack, D; Riedewald, J; Rohde, H; Magnus, T; Feucht, HH; Elsner, HA, et al. Essential functional role of the polysaccharide intercellular adhesin of *Staphylococcus epidermidis* in hemagglutination. *Infect Immun.*, 1999, 67(2), 1004-8.

[15] Vuong, C; Kocianova, S; Voyich, JM; Yao, Y; Fischer, ER; DeLeo, FR; et al. A crucial role for exopolysaccharide modification in bacterial biofilm formation, immune evasion, and virulence. *J Biol Chem.*, 2004, 279(52), 54881-6.

[16] Fluckiger, U; Ulrich, M; Steinhuber, A; Döring, G; Mack, D; Landmann, R; et al. Biofilm formation, *icaADBC* transcription, and polysaccharide intercellular adhesin synthesis by staphylococci in a device-related infection model. *Infect Immun.*, 2005, 73(3), 1811-9.

[17] Christensen, GD; Simpson, WA; Bisno, AL; Beachey EH. Adherence of slime-producing strains of *Staphylococcus epidermidis* to smooth surfaces. *Infect Immun.*, 1982, 37(1), 318-26.

[18] Hall, RT; Hall, SL; Barnes, WG; Izuegbu, J; Rogolsky, M; Zorbas I. Characteristics of coagulase-negative Staphylococci from infants with bacteremia. *Pediatr Infect Dis J.*, 1987, 6(4), 377-83.

[19] Christensen, GD; Parisi, JT; Bisno, AL; Simpson, WA; Beachey EH. Characterization of clinically significant strains of coagulase-negative staphylococci. *J Clin Microbiol.*, 1983, 18(2), 258-69.

[20] Kotilainen P. Association of coagulase-negative staphylococcal slime production and adherence with the development and outcome of adult septicemias. *J Clin Microbiol.*, 1990, 28(12), 2779-85.

[21] Riley, TV; Schneider PF. Infrequency of slime production by urinary isolates of *Staphylococcus saprophyticus*. *J Infect.*, 1992, 24(1), 63-6.

[22] Cunha, MLRS; Rugolo, LMSS. Lopes CAM. Study of virulence factors in coagulase-negative staphylococci isolated from newborns. *Mem Inst Oswaldo Cruz.*, 2006, 101, 661-8.

[23] Cunha, MLRS; Caramori, JCT; Fioravante, AM:, Batalha, JEN; Montelli, AC; Barretti, P. The significance of slime as virulence factor in coagulase-negative staphylococci in peritonitis. *Perit Dial Intern.*, 2004, 24, 191-3.

[24] Freeman, DJ; Falkiner, FR; Keane, CT. New method for detecting slime production by coagulase negative Staphylococci. *J Clin Pathol.*, 1989, 42(8), 872-4.

[25] Christensen, GD; Simpson, WA; Younger, JJ; Baddour, LM; Barrett, FF; Melton, DM; et al. Adherence of coagulase-negative Staphylococci to plastic tissue culture plates: a quantitative model for the adherence of

staphylococci to medical devices. *J Clin Microbiol.*, 1985, 22(6), 996-1006.

[26] Oliveira, A; Cunha, MLRS. Comparison of methods for the detection of biofilm production in coagulase-negative staphylococci. *BMC Res Notes.*, 2010, 3, 260.

[27] Arciola, CR; Baldassarri, L; Montanaro, L. Presence of *icaA* and *icaD* genes and slime production in a collection of staphylococcal strains from catheter-associated infections. *J Clin Microbiol.*, 2001, 39(6), 2151-6.

[28] Davenport, DS; Massanari, RM; Pfaller, MA; Bale, MJ; Streed, SA; Hierholzer, WJ. Usefulness of a test for slime production as a marker for clinically significant infections with coagulase-negative Staphylococci. *J Infect Dis.*, 1986, 153(2), 332-9.

[29] Davey, ME; O'Toole, GA. Microbial biofilms: from ecology to molecular genetics. *Microbiol Mol Biol Rev.*, 2000, 64(4), 847-67

[30] Hussain, M; Herrmann, M; von, Eiff, C; Perdreau-Remington, F; Peters, G. A 140-kilodalton extracellular protein is essential for the accumulation of *Staphylococcusepidermidis* strains on surfaces. *Infect Immun.*, 1997, 65(2), 519-24.

[31] Rohde, H; Burandt, EC; Siemssen, N; Frommelt, L; Burdelski, C; Wurster, S; et al. Polysaccharide intercellular adhesin or protein factors in biofilm accumulation of *Staphylococcus epidermidis* and *Staphylococcusaureus* isolated from prosthetic hip and knee joint infections. *Biomaterials.*, 2007, 28(9), 1711-20.

[32] Rohde, H; Burdelski, C; Bartscht, K; Hussain, M; Buck, F; Horstkotte, MA; et al. Induction of *Staphylococcus epidermidis* biofilm formation via proteolytic processing of the accumulation-associated protein by staphylococcal and host proteases. *Mol Microbiol.*, 2005, 55(6), 1883-95.

[33] Cucarella, C; Solano, C; Valle, J; Amorena, B; Lasa, I; Penadés, JR. Bap, a *Staphylococcus aureus* surface protein involved in biofilm formation. *J Bacteriol.*, 2001, 183(9), 2888-96.

[34] Gill, SR; Fouts, DE; Archer, GL; Mongodin, EF; Deboy, RT; Ravel, J; et al. Insights on evolution of virulence and resistance from the complete genome analysis of an early methicillin-resistant *Staphylococcus aureus* strain and a biofilm-producing methicillin-resistant *Staphylococcus epidermidis* strain. *J Bacteriol.*, 2005, 187(7), 2426-38.

[35] Izano, EA; Amarante, MA; Kher, WB; Kaplan, JB. Differential roles of poly-N-acetylglucosamine surface polysaccharide and extracellular

DNA in *Staphylococcus aureus* and *Staphylococcus epidermidis* biofilms. *Appl Environ Microbiol.*, 2008, 74(2), 470-6.
[36] Allesen-Holm, M; Barken, KB; Yang, L; Klausen, M; Webb, JS; Kjelleberg, S; et al. A characterization of DNA release in *Pseudomonas aeruginosa* cultures and biofilms. *Mol Microbiol.*, 2006, 59(4), 1114-28.
[37] Qin, Z; Yang, X; Yang, L; Jiang, J; Ou, Y; Molin, S; et al. Formation and properties of in vitro biofilms of *ica*-negative *Staphylococcus epidermidis* clinical isolates. *J Med Microbiol.*, 2007, 56(Pt 1), 83-93.
[38] Mann, EE; Rice, KC; Boles, BR; Endres, JL; Ranjit, D; Chandramohan, L; et al. Modulation of eDNA release and degradation affects *Staphylococcus aureus* biofilm maturation. *PLoS One.*, 2009, 4(6), e5822
[39] March, JC; Bentley, WE. Quorum sensing and bacterial cross-talk in biotechnology. *Curr Opin Biotechnol.*, 2004, 15(5), 495-502.
[40] Korem, M; Gov, Y; Kiran, MD; Balaban, N. Transcriptional profiling of target of RNAIII-activating protein, a master regulator of staphylococcal virulence. *Infect Immun.*, 2005, 73(10), 6220-8.
[41] Gov, Y; Borovok, I; Korem, M; Singh, VK; Jayaswal, RK; Wilkinson, BJ; et al. Quorum sensing in Staphylococci is regulated via phosphorylation of three conserved histidine residues. *J Biol Chem.*, 2004, 279(15), 14665-72.
[42] Balaban, N; Goldkorn, T; Gov, Y; Hirshberg, M; Koyfman, N; Matthews, HR; et al. Regulation of *Staphylococcus aureus* pathogenesis via target of RNAIII-activating Protein (TRAP). *J Biol Chem.*, 2001, 276(4), 2658-67.
[43] Novick, RP; Ross, HF; Projan, SJ; Kornblum, J; Kreiswirth, B; Moghazeh, S. Synthesis of staphylococcal virulence factors is controlled by a regulatory RNA molecule. *EMBO J.*, 1993, 12(10), 3967-75.
[44] Balaban, N; Stoodley, P; Fux, CA; Wilson, S; Costerton, JW; Dell'Acqua, G. Prevention of staphylococcal biofilm-associated infections by the quorum sensing inhibitor RIP. *Clin Orthop Relat Res.*, 2005(437), 48-54.
[45] Yang, G; Gao, Y; Dong, J; Liu, C; Xue, Y; Fan, M; et al. A novel peptide screened by phage display can mimic TRAP antigen epitope against *Staphylococcus aureus* infections. *J Biol Chem.*, 2005, 280(29), 27431-5.
[46] Balaban, N; Giacometti, A; Cirioni, O; Gov, Y; Ghiselli, R; Mocchegiani, F; et al. Use of the quorum-sensing inhibitor RNAIII-

inhibiting peptide to prevent biofilm formation in vivo by drug-resistant *Staphylococcus epidermidis*. *J Infect Dis.*, 2003, 187(4), 625-30.

[47] Balaban, N; Cirioni, O; Giacometti, A; Ghiselli, R; Braunstein, JB; Silvestri, C; et al. Treatment of *Staphylococcus aureus* biofilm infection by the quorum-sensing inhibitor RIP. *Antimicrob Agents Chemother.*, 2007, 51(6), 2226-9.

[48] Simonetti, O; Cirioni, O; Ghiselli, R; Goteri, G; Scalise, A; Orlando, F; et al. RNAIII-inhibiting peptide enhances healing of wounds infected with methicillin-resistant *Staphylococcus aureus*. *Antimicrob Agents Chemother.*, 2008, 52(6), 2205-11.

[49] Sauer, K; Camper, AK; Ehrlich, GD; Costerton, JW; Davies, DG. *Pseudomonas aeruginosa* displays multiple phenotypes during development as a biofilm. *J Bacteriol.*, 2002, 184(4), 1140-54.

[50] Drenkard, E. Antimicrobial resistance of *Pseudomonas aeruginosa* biofilms. *Microbes Infect.*, 2003, 5(13), 1213-9.

[51] Plouffe, JF. Emerging therapies for serious gram-positive bacterial infections: a focus on linezolid. *Clin Infect Dis.*, 2000, 31 Suppl 4, S144-9.

[52] Whiteley, M; Bangera, MG; Bumgarner, RE; Parsek, MR; Teitzel, GM; Lory, S; Greenberg, EP. Gene expression in *Pseudomonas aeruginosa* biofilms. *Nature.*, 2001, 413(6858), 860-4.

Chapter 6

STAPHYLOCOCCAL TOXINS

Several virulence factors are responsible for the symptoms and severity of infections caused by *Staphylococcus* spp. These factors include alpha-, beta-, gamma- and delta-toxin, Panton-Valentine leukocidin, and the group of pyrogenic toxin superantigens (PTSAgs) [1].

Staphylococcal toxins can be generally divided into two groups: membrane-active agents that include alpha-, beta-, delta- and gamma-toxin and leukocidin, and toxins with superantigen activity that include the family of PTSAgs. These toxins comprise staphylococcal enterotoxins, toxic shock syndrome toxin 1 (TSST-1), and the family of exfoliative toxins [2]. Some of the genes encoding these toxins are frequently located on mobile genetic elements, such as phages and *S. aureus* pathogenicity islands (SaPIs), which are potentially mobile DNA segments of variable size that encode virulence-associated genes and are transferred horizontally between strains. Examples are the genes encoding staphylococcal enterotoxin B (SEB) and C (SEC) and TSST [3].

According to Peacock et al. [4], the number of virulence-associated genes carried by a bacterial strain is the product of the interaction between the rates of gene acquisition, the biological fitness cost, and the rate of decay of the disease-causing strain. Since the vast majority of severe infections caused by *Staphylococcus* spp. cannot be explained by the action of a single virulence factor, it is likely that several factors act together during the infective process.

In addition to biofilms, *Staphylococcus* spp. produce other virulence factors such as hemolysins, lipases, proteases and toxins which may also play a yet unknown role in the pathogenesis of severe infections. At least three cytolytic or membrane-damaging toxins are produced by coagulase-negative staphylococci (CoNS), including alpha-, beta- and delta-toxin (5-8). These toxins have been described as hemolysins based on their ability to lyse red blood cells, but since their biological activity is not restricted to red blood cells, the term cytolytic toxin was introduced [9].

Among these toxins, a hemolysin resembling *S. aureus* delta-toxin has received attention from researchers due to its involvement in necrotizing enterocolitis that affects newborns [6-8, 10]. Several authors have described the occurrence of necrotizing enterocolitis in association with CoNS [11, 12]. This delta-toxin can damage a variety of cells due to its detergent-like action on the cell membrane and is distinguished from other hemolysins by its thermostability and neutralization by lecithin [13]. Furthermore, studies indicate that the delta-toxin plays an important role in the pathogenesis of intestinal diseases, ranging from acute diarrhea to severe enteritis [10, 13].

The production of alpha-hemolysin is also associated with the pathogenicity of *Staphylococcus* spp. due to the known ability of this toxin to damage tissues after establishment of a focus of infection [14] and its dermonecrotic and lethal activity in experimental animals [15]. Stephen and Pietrowski [13] observed variation in the susceptibility of erythrocytes of different species to alpha-hemolysin, with rabbit erythrocytes being 1,000 times more susceptibility than human cells.

A study conducted by our group which compared the evolution of peritonitis caused by *S. aureus* and CoNS in peritoneal dialysis patients from the Dialysis Unit of the University Hospital, Botucatu Medical School (HC-FMB) [16], showed that the production of alpha-hemolysin was related to poor outcome of peritonitis, with the rate of resolution being 8.2 times higher for peritonitis caused by *Staphylococcus* spp. not producing this toxin. These results agree with the data published by Haslinger-Löffler et al. [17] who suggested that alpha-hemolysin plays a specific role in the pathogenesis of peritonitis. These authors showed that only invasive and alpha-hemolysin-producing strains of *S. aureus* induced caspase-independent cell death in mesothelial cells. These results indicate that alpha-hemolysin represents an important mechanism of *S. aureus* to cause persistent damage to the peritoneum during peritonitis. In contrast to *S. aureus*, no cytotoxic effects were observed for the *S. epidermidis* strains tested, which were not invasive and did not produce alpha-hemolysin.

Recently, another study conducted in the same Dialysis Unit, but analyzing only episodes of peritonitis caused by *S. aureus* over a period of 15 years (1996 to 2010), indicated the production of beta-hemolysin as a factor that was independently associated with the lack of resolution of peritonitis

[18]. Beta-hemolysin, also called sphingomyelinase C, is a thermolabile protein that is toxic to different cells, including erythrocytes, leukocytes and macrophages [13]. This enzyme catalyzes the hydrolysis of membrane phospholipids in susceptible cells, with the lysis intensity being proportional to the concentration of sphingomyelin present in the cytoplasmic membrane exposed on the cell surface [14]. The incubation of small amounts of beta-hemolysin with sheep erythrocytes at 37°C results in little or no lysis. However, the hemolytic property of the toxin is enhanced by subsequent exposure of erythrocytes to a temperature of 4°C, a phenomenon called hot-cold hemolysis.

Beta-hemolysin does not lyse most types of host cells, but leaves them vulnerable to a series of other lytic agents, such as alpha-hemolysin and Panton-Valentine leukocidin [19]. In a model of *S. aureus*-induced lung injury in mice, Hayashida et al. [19] observed that lung injury was significantly attenuated in animals infected with *S. aureus* deficient in beta-hemolysin compared to animals infected with *S. aureus* expressing this toxin. This experimental disease was characterized by intense neutrophil inflammation and a reduction in the expression of syndecan-1 in alveolar epithelial cells and could be reproduced by the administration of recombinant beta-hemolysin, but not by beta-hemolysin mutants deficient in sphingomyelinase activity.

Recently, Huseby et al. (20) demonstrated that beta-hemolysin forms cross-links with extracellular DNA, independent of sphingomyelinase activity, producing an insoluble nucleoprotein *in vitro* and contributing to biofilm formation. Using a rabbit model of infectious endocarditis, the authors observed that this toxin stimulates biofilm formation *in vivo*.

Another important cytotoxin that has attracted the attention of many researchers is Panton-Valentine leukocidin (PVL), encoded by the *lukS-PV* and *lukF-PV* genes, which is found mainly in community-acquired methicillin-resistant *S. aureus* (CA-MRSA). PVL causes extensive tissue necrosis and is responsible for severe skin infections and necrotizing pneumonia. Leukocidin has been associated with skin and soft tissue infections by two researchers, Panton and Valentine, in 1932. The *lukS-PV* and *lukF-PV*genes are carried by a specific bacteriophage (phiSLT) which infects and lysogenizes *S. aureus* cells, integrating the PVL genes into the chromosome of this microorganism. Once the LukS-PV and LukF-PV proteins are transcribed and secreted, they assemble into a pore-forming heptamer on the membrane of polymorphonuclear leukocytes (PMNs), causing lysis of these cells. Depending on the concentration of PVL, PMNs undergo either lysis or apoptosis. In view of this evidence, the action of PVL is probably not directly associated with tissue necrosis, but rather with the release of cytotoxic lysosomal granules from lysed PMNs, the release of reactive oxygen species by granulocytes, or the inflammatory cascade [21].

A study conducted in our laboratory using strains isolated from clinical and surveillance cultures of patients seen at Hospital Estadual Bauru [22] showed that 18.1% of methicillin-resistant *S. aureus* (MRSA) strains were positive for the *pvl* gene versus 7.5% of methicillin-sensitive *S. aureus* (MSSA) ($p < 0.009$). Although several authors reported the production of this toxin to be more frequent among MRSA, results obtained in another study from our group using *S. aureus* isolated from skin infections showed that all isolates carrying the PVL gene were susceptible to oxacillin and none of the CA-MRSA strains carried the PVL gene [23]. These results agree with the model of Boyle-Vavra and Daum [21] which suggests that an oxacillin-sensitive *S. aureus* strain is first infected and lysogenized by a phage that carries the PVL genes and later acquires the *mecA* gene by horizontal transfer.

In addition to cytotoxins, staphylococci can also produce toxins with superantigen activity. This group includes enterotoxins, hydrosoluble exoproteins with a molecular weight of 26,000 to 29,000 Da that are rich in lysin, aspartic acid and glutamic acid and contain two cysteines forming a disulfide bridge [24]. Enterotoxins are relatively resistant to heat and the proteolytic enzymes trypsin, pepsin, renin and papain, a fact permitting their passage through the gastrointestinal tract without loss of activity [25].

Staphylococcal enterotoxins began to be purified and characterized in 1959 [26]. To date, 23 serologically distinct enterotoxins have been described, which are designated SEA to SEIV [27, 28]. All enterotoxins exhibit superantigen activity, but only some (SEA to SEI, SER, SES and SET) have been proven to be emetic [29, 30]. According to the International Nomenclature Committee for Staphylococcal Superantigens, only superantigens that induce emesis after oral administration in an experimental primate model should be designated staphylococcal enterotoxins. The Committee also recommends that other similar toxins that do not present emetic properties in primate models or that have not yet been tested should be designated staphylococcal enterotoxin-like (SEl) toxin type X [31, 32].

A toxin involved in toxic shock syndrome (TSS) was first designated staphylococcal enterotoxin F (SEF) [24]. However, this toxin did not exhibit emetic activity *in vivo* after oral administration, a characteristic of a true enterotoxin, and was therefore later renamed TSST-1 [33]. All genes encoding staphylococcal toxins are located on mobile genetic elements, including bacteriophages, SaPIs, and plasmids [34-36]. The genes encoding enterotoxins

SEB, SEC, SEG, SEI, SElM, SElN, SElO, SElK, SElQ, and TSST-1 are located on pathogenicity islands [37-41]. The SEA, SEE and SElP genes are carried by prophages [39, 42, 43], whereas the SED, SElJ and SER genes are found on plasmid pIB485 [44-46]. Among the four pathogenicity islands described in *S. aureus* (SaPIs), SaPI3 is important since it forms a cluster of staphylococcal enterotoxin genes, called enterotoxin gene cluster (*egc*), which contains the genes encoding SEG (*seg*), SEI (*sei*), SElM (*selm*), SElN (*seln*) and SElO (*selo*), in addition to two pseudogenes (*Φent1* and *Φent2*) of still unknown biological function [38, 47]. The horizontal transfer of enterotoxin genes between staphylococcal strains plays an important role in the evolution of *S. aureus* and CoNS as pathogens. Knowledge of the true extent of the diversity of staphylococcal superantigens is important for a better understanding of the pathogenicity or virulence of staphylococci [46].

The detection of staphylococcal enterotoxin is decisive to confirm an outbreak of food poisoning and to determine the enterotoxigenicity of strains [48]. Agar-gel immunodiffusion tests were the first methods developed and included single diffusion, one-dimensional diffusion, double diffusion, microslide technique, and capillary tube diffusion [48]. However, doubts have been raised regarding the sensitivity of gel diffusion methods for enterotoxin detection. Significant proportions of toxigenic strains produce concentrations of 1 ng/ml or less, which are below the detection levels of diffusion methods [49].

In this respect, methods such as reverse passive hemagglutination, reverse passive latex agglutination (RPLA), radioimmunoassay and enzyme-linked immunosorbent assay (ELISA) have been developed whose sensitivity ranges from 1.0 to 0.1 ng/ml [48, 50]. However, nanogram detection of enterotoxins can be impaired by false-positive reactions resulting from the interference of food components, microbial antigens and extracellular metabolites, compromising already available methods [48, 50, 51].

Furthermore, commercial antiserum is available only for SEA, SEB, SEC, SED, and SEE. Experimental tests have been developed for some of the new toxins (SEG, SEH and SEI), but are not commercially available due to difficulties in the purification and preparation of specific antibodies [52]. A variety of ELISA-based methods for the detection of staphylococcal enterotoxins have been described [53], including the automated VIDAS Staph Enterotoxin II (SET2) system (BioMerieux), Tecra™ kit (3M), and Elisa Transia™ kit (Transia-Diffchamb S.A., Lyon, France). In Brazil, the RPLA kit from Oxoid has been frequently used for the detection of enterotoxins in foods and culture supernatants [8, 51, 54]. These kits are commonly employed because of their simplicity and sensitivity, but may yield false-positive results due to crossreactivity and the occurrence of nonspecific reactions.

In an attempt to overcome this problem, amplification techniques such as the polymerase chain reaction (PCR) have been developed, which permit the identification of the genes responsible for the production of enterotoxins and TSST-1, with high sensitivity and specificity. Johnson et al. [55] developed a protocol for detecting the genes encoding SEA to SEE and TSST-1 using oligonucleotides synthesized based on the computer analysis of the previously published sequences of the staphylococcal enterotoxin genes (*sea*, *seb*, *sec-1*, *sed*, and *see*) and *tst*.

Despite the scant attention given to the toxigenic profile of CoNS, some researchers emphasize that these microorganisms can produce TSST-1 alone or in combination with one of the staphylococcal enterotoxins, and their clinical importance and toxigenic capacity should therefore not be ignored [56-58]. In the study of Crass and Bergdoll [56], strains producing TSST-1 alone or TSST-1 and SEA were isolated from seven (50%) of 14 women with TSS. Furthermore, CoNS were isolated from four patients with TSS who had an *S. aureus*-positive culture. However, only the CoNS isolates produced the toxin. CoNS strains producing TSST-1, SEA and SEC were also isolated from patients with other infections and from food implicated in a case of food poisoning. Similar results have been reported by Kahler et al. [57] who described a case of TSS in which the TSST-1-producing strain was isolated from the patient's vagina, whereas no *S. aureus* isolate was found. Valle et al. [59], studying the production of TSST-1 by staphylococcal strains isolated from goats, observed that 16.5% of the CoNS isolates were producers of this toxin.

Several studies using PCR for the confirmation of the enterotoxigenicity of staphylococci have found divergences between the results obtained with immunological assays and PCR [55, 60], indicating possible errors due to the occurrence of false-positive results. In this respect, studies have been conducted in our laboratory to demonstrate the enterotoxigenic capacity of CoNS. In the first study [61], the main objective was to detect enterotoxin genes in CoNS isolated from foods. Among the 20 CoNS isolates tested, three carried the *sea* gene and one the *sec-1* gene. The *sea* gene was detected in one *S. epidermidis* isolate, one *S. xylosus* isolate and one *S. hominis* isolate, whereas the *sec-1* gene was detected in one *S. xylosus* isolate. However, none of these toxins was identified by the RPLA method.

Udo et al. [62], studying restaurant workers, identified CoNS and *S. aureus* strains producing enterotoxins and TSST-1 by the RPLA method.

Among the CoNS isolates studied, 14.1% were producers of enterotoxins or TSST-1. Strains of *S. hominis*, *S. warneri*, *S. saprophyticus*, *S. epidermidis*, *S. xylosus*, *S. haemolyticus* and *S. schleiferi* were positive for SEA, SEB, SEC, and/or TSST-1. Marín et al. [63] studied enterotoxin-producing staphylococci isolated from dry-cured ham by the RPLA method. Two of the 135 staphylococcal isolates belonged to the species *S. epidermidis* and one was a producer of SEC. In the study of Johnson et al. [55], the phenotype determined by RPLA differed from the genotype observed by PCR in two of 88 *S. aureus* strains. These strains carried the *sec* and *tst* genes, but production of these toxins was not detected by RPLA. Schmitz et al. [64] evaluated the presence of the *sec*, *seb* and *tst* genes in 50 *S. aureus* isolates and identified two strains that carried the *sec* gene, but did not produce SEC when tested by RPLA.

In a second study from our group [65], PCR was used to detect the genes responsible for the production of enterotoxins and TSST-1 in 120 *S. aureus* and 120 CoNS strains isolated from newborns. The results obtained were compared to the detection of enterotoxins by the RPLA method. Among the 120 *S. aureus* isolates, 38.3% were identified as enterotoxin producers by RPLA, whereas PCR detected the genes in 46.6% of the strains. Similar results were observed for the CoNS strains, with the frequency of detection of specific genes by PCR (40%) being higher than the frequency of enterotoxin production demonstrated by the RPLA method (26.7%).

However, other authors were unable to detect the presence of enterotoxin-encoding genes in clinical isolates of CoNS, such as Becker et al. [66] using multiplex PCR. Kreiswirth et al. [67] evaluated the presence of genes responsible for TSST-1 production in different CoNS species using DNA hybridization. The authors included in their study some of the strains reported by Crass and Bergdoll [56] to be TSST-1 positive. The presence of the *tst* gene was not confirmed in any of the CoNS strains. Additionally, two TSST-1-positive isolates, previously identified as coagulase negative, were coagulase positive, indicating an identification error.

Other authors also questioned the ability of CoNS to produce enterotoxins and TSST-1 and suggested these strains to be *S. aureus* mutants that do not express the enzyme coagulase [68, 69]. To rule out this possibility, in our laboratory all CoNS isolates that carry toxin genes are submitted to genotypic identification for species confirmation. In this respect, ITS-PCR leaves no doubt regarding the characteristic band pattern of each species analyzed when compared to international reference strains (ATCC), in addition to ruling out possible phenotypic identification errors. Despite some discrepancies between methods, in terms of species differentiation, all isolates identified as CoNS by the phenotypic method were confirmed by the genotypic technique.

Eliminating possible identification errors, the divergence in the results obtained regarding the ability of CoNS to produce toxins or not might be related to the choice of the adequate technique for DNA extraction and for enterotoxin gene detection. The standardization of techniques that ranges from the ideal culture medium, pH, temperature and nutrient concentration to the use of appropriate extraction controls, amplification parameters and enzymes is fundamental for good performance of the research. One disadvantage of multiplex PCR is the fact that it highlights genes present at higher frequency, masking or not detecting genes that appear at a lower frequency. In contrast, PCR uses one primer pair for each reaction, which is performed separately for each toxin studied.

PCR detects genes of interest independently of their expression; thus, genes responsible for enterotoxin production may be present but are not active. The reverse transcriptase-polymerase chain reaction (RT-PCR) is a molecular technique that permits the detection of messenger RNA (mRNA). In this method, mRNA is converted to complementary DNA (cDNA) by the enzyme reverse transcriptase, with the expression of the gene being proportional to the number of mRNA copies of the gene of interest. The cDNA is amplified by PCR using specific primers to confirm expression, which leaves no doubt regarding the toxigenic potential of the microorganism. In a subsequent study from our group using RT-PCR for the detection of SEA, SEB, SEC, SED and TSST-1 in *S. aureus* and CoNS isolates [70], expression of toxin mRNA was observed in 43 (39.8%) of 108 PCR-positive strains. Toxin expression was significantly lower in CoNS than in *S. aureus*, corresponding to 13.9% of all isolates with a toxigenic potential. *Staphylococcus epidermidis* was the most toxigenic CoNS species, with the observation of expression of SEA and SEC mRNA in five isolates. Among the other CoNS species, only *S. lugdunensis* presented a positive RT-PCR result for SEC.

RT-PCR was also used for the detection of mRNA encoding enterotoxins SEE, SEG, SEH and SEI in 90 *S. aureus* isolates and in 90 CoNS isolates [71]. mRNA expression was observed in 42 (50.6%) of the 83 PCR-positive isolates. The mRNA detected corresponded to toxins of the SEG, SEH and SEI classes. Toxin expression was lower in CoNS than in *S. aureus*, corresponding to 34.5% of all producing isolates versus 59.3% for *S. aureus*. *Staphylococcus epidermidis* was the most toxigenic CoNS species. Expression of SEG, SEH and SEI mRNA was observed in seven *S. epidermidis* isolates. Among the other CoNS species, only *S. warneri* and *S. lugdunensis* presented a positive RT-PCR, with the former being positive for SEI and the latter for SEG and SEI.

Some associations found by PCR call attention since they exist in various strains, such as the concomitant presence of the *seg* and *sei* genes in 16 *S. aureus* isolates and in 10 CoNS isolates. The coexistence of the *seg* and *sei* genes might be due to their chromosomal location, with both genes belonging to an *egc* operon which contains five enterotoxin genes (*seg, sei, selm, seln* and *selo*) and is located in SaPIs. Maiques et al. [72] reported the transfer of an SaPI by a bacteriophage from an *S. aureus* strain to another staphylococcal species such as *S. epidermidis*.

Finally, our results regarding the ability of CoNS strains to express staphylococcal superantigens are consistent with the study of Madhusoodanan et al. [73] who provided the first evidence of a pathogenicity island in a clinical isolate of *S. epidermidis* (SePI). Sequencing and analysis of the genome of this strain demonstrated the presence of SePI-1 which encodes the SEC-3 and SElL genes. Furthermore, the expression of these superantigens was confirmed by quantitative RT-PCR and Western blotting.

RT-PCR has been shown to be a rapid and efficient technique, demonstrating the ability of CoNS to express mRNA encoding these enterotoxins. However, since toxin production depends on the activation of genes, further investigation of environmental factors and elucidation of the regulatory mechanisms that interfere with their expression are needed.

The expression of genes encoding virulence factors in *S. aureus* is coordinated by global regulators. These regulators help bacteria to adapt to a hostile environment by the production of factors that enable them to survive and to subsequently cause infection at the appropriate time. Several of the global virulence regulators, such as the *agr* (accessory gene regulator), *sar*(staphylococcal accessory regulator) and *sae* (staphylococcal accessory element) systems, have been characterized [74]. Other systems such as *arl* (autolysis-related locus), *sar* homologs (*rot* – repressor of toxins, *mgrA* – multiple global regulator, *sarS*, *saR*, *sarT*, *sarU*, *sarV*, *sarX*, *sarZ* and *tcaR* – teicoplanin-associated locus regulator), the *srr* system (staphylococcal respiratory response), and *trap* (target of the RNA III-activating protein) require further investigation to determine their exact roles in the regulation of virulence [74]. The best characterized regulatory system is the *agr* system. According to Novick and Jiang [75], expression of the *agr* locus is an important regulator of many virulence factors in *S. aureus*.

The *agr* locus comprises a cluster of genes with quorum sensing activity. Quorum sensing is a mechanism whereby the bacterium is able to "perceive" the cell density in its environment through cell-cell communication, permitting a phenotypic reaction according to the growth phase of the culture. Quorum sensing is important in *Staphylococcus* since certain proteins, such as virulence factors, should only be expressed in a certain phase of growth [75]. This system coordinates the expression of genes related to the biological requirements of *Staphylococcus*, permitting adhesion to host cells and tissue, dissemination in the organism, and the degradation of cells and tissues for both nutrition and protection against the host defense [76].

The expression of the *agr* system contributes to staphylococcal pathogenicity at different times during infection. Analysis of the growth curve of *Staphylococcus* shows that, except for enterotoxin A which is produced throughout all phases of growth, all exoproteins are secreted during the post-exponential phase. On the other hand, membrane-associated proteins and adhesins are produced during the exponential phase and not the post-exponential phase [77].

Four polymorphisms have been described in the *agr* locus of *Staphylococcus*, which are called *agr* I, II, III and IV [78]. In these groups, variations occur in the *agrB*, *agrC* and a*grD* genes and, consequently, in AIP and protein AgrC. The binding of AIP to its receptor is specific for each allelic group. When an AIP of one allelic group binds to an AgrC receptor of another group, AIP does not produce intrinsic factor and no signal transduction through AgrC occurs. In this case, AIP functions as an antagonist. An AIP only acts as an agonist for its own allelic group, so that bacteria of different *agr* groups interfere with the regulation of accessory proteins of one another [79].

Novick et al. [75] described the expression of the *agr* locus as an important regulator of many virulence factors in *S. aureus*. Although several authors question the toxigenic potential of CoNS [68, 69], the *agr* operon, which plays an important role in the regulation of staphylococcal toxin expression, has also been found in other staphylococcal species such as *S. intermedius* [80], *S. lugdunensis* [81] and *S. epidermidis* [82]. The predominance of *S. epidermidis* and the detection of *S. lugdunensis* among toxigenic CoNS in our study may be related to the fact that these staphylococcal regulatory systems are also present in these species.

A novel class of amphipathic peptides, called phenol-soluble modulins (PSM), are acquiring a major role in the virulence of *Staphylococcus* spp. [83].

In contrast to most other toxins, PSM are small peptides encoded by chromosomal genes, except for PSM-*mec* which is located on a mobile genetic element (staphylococcal cassette chromosome *mec*, SCC*mec*) that links resistance and virulence genes on the same genetic element [84]. The production of PSM is controlled by quorum sensing (*agr*). These peptides exert proinflammatory activity and exhibit a high capacity to recruit, stimulate and lyse human neutrophils, significantly contributing to the pathogenesis of infection [85].

Despite their similar structure, PSM are classified into two groups depending on their size. Alpha type PSM (PSMα1, PSMα2, PSMα3, PSMα4, and δ-toxin) are smaller, with a length of 20-30 amino acids, and are considered the most toxic [85], whereas beta type PSM (PSMβ1 and PSMβ2) consist of approximately 44 amino acids and seem to exert additional functions. For example, the beta type PSM of *S. epidermidis* have been described to play a role in biofilm dispersion [86].

In view of their amphipathic nature, PSM are important for the formation of biofilm channels which are essential for the passage of nutrients into this structure and permit the release of bacteria present in the biofilm and the consequent dissemination of infection to other sites of the body. PSM are found in practically all *Staphylococcus* spp., especially pathogenic strains [87].

It appears that staphylococci can use combinations of different virulence factors which result in clinical syndromes and possibly contribute to the dispersion of bacteria in the host organism. Superantigens are able to stimulate the activity of T lymphocytes, inducing the production of high levels of cytokines that can cause anergy, inflammation, cytotoxicity, T-cell depletion and autoimmunity [27, 88], events that facilitate colonization of the individual.

Commensal or pathogenic *Staphylococcus* species possess a wide array of virulence factors that are expressed or not, depending on the stage of the life cycle of the organism, the environment, and the existing regulatory systems that can directly affect toxin production. These factors, which are responsible for direct tissue invasion or toxigenicity due to the activity of secreted toxins, may explain the high prevalence and success of these microorganisms in community-acquired and nosocomial infections. The molecular basis of staphylococcal pathogenicity is multifactorial, depending on the presence and also on the expression of various accessory genes. Therefore, the presence of toxin genes in staphylococcal strains in the absence of mRNA transcription of the gene does not rule out the possibility of toxin production at a given time point, either *in vivo* or in culture media under optimal conditions.

REFERENCES

[1] Koneman, EW; Allen, SD; Janda, WM; Schreckenberger, PC. Color Atlas and Textbook of Diagnostic Microbiology. 5th ed. Philadelphia: J.B. Lippincott: 1997.

[2] Bohach, GA; Fast, DJ; Nelson, RD; Schlievert, PM. Staphylococcal and streptococcal pyrogenic toxins involved in toxic shock syndrome and related illnesses. *Crit Rev Microbiol.*, 1990, 17(4), 251-72.

[3] Hanssen, AM; Ericson, Sollid, JU. SCC*mec* in Staphylococci: genes on the move. *FEMS Immunol Med Microbiol.*, 2006, 46(1), 8-20.

[4] Peacock, SJ; Moore, CE; Justice, A; Kantzanou, M; Story, L; Mackie, K; et al. Virulent combinations of adhesin and toxin genes in natural populations of *Staphylococcus aureus*. *Infect Immun.*, 2002, 70(9), 4987-96.

[5] Lambe, DW; Ferguson, KP; Keplinger, JL; Gemmell, CG; Kalbfleisch, JH. Pathogenicity of *Staphylococcus lugdunensis*, *Staphylococcus schleiferi*, and three other coagulase-negative Staphylococci in a mouse model and possible virulence factors. *Can J Microbiol.*, 1990, 36(7), 455-63.

[6] Molnàr, C; Hevessy, Z; Rozgonyi, F; Gemmell, CG. Pathogenicity and virulence of coagulase negative Staphylococci in relation to adherence, hydrophobicity, and toxin production in vitro. *J Clin Pathol.*, 1994, 47(8), 743-8.

[7] Marconi, C; Cunha, MLRS; Araújo, JR, JP; Rugolo, LMSS, Standardization of the PCR technique for the detection of delta toxin in *Staphylococcus* spp. *J. Venom. Anim. Toxins incl. Trop. Dis.*, 2005,11 (2),117-128.

[8] Cunha, MLRS; Rugolo, LM; Lopes, CA. Study of virulence factors in coagulase-negative Staphylococci isolated from newborns. *Mem Inst Oswaldo Cruz.*, 2006, 101(6), 661-8.

[9] Bernheimer, AW. Interactions between membranes and cytolytic bacterial toxins. *Biochim Biophys Acta.*, 1974,344, 27-50.

[10] Scheifele, DW; Bjornson, GL; Dyer, RA; Dimmick, JE. Delta-like toxin produced by coagulase-negative Staphylococci is associated with neonatal necrotizing enterocolitis. *Infect Immun.*, 1987, 55(9), 2268-73.

[11] Rotbart, HA; Johnson, ZT; Reller, LB. Analysis of enteric coagulase-negative Staphylococci from neonates with necrotizing enterocolitis. *Pediatr Infect Dis J.*, 1989, 8(3), 140-2.

[12] Dalal, A; Urban, C. Enterocolitis caused by methicillin-resistant *Staphylococcus aureus*. *Infect Dis Clin Prac.*, 2008, 16(4), 222-223.
[13] Stephen, J; Pietrowski, RA. *Bacterial Toxins*. 2nd ed. Wokingham, England: van Nostrand Reinhold: 1986.
[14] Dinges, MM; Orwin, PM; Schlievert, PM. Exotoxins of *Staphylococcus aureus*. *Clin Microbiol Rev.*, 2000,13(1), 16-34.
[15] Bernheimer, AW; Schwartz. Isolation and composition of staphylococcal alpha toxin. *J. Gen. Microbiol.*, 1963, 30, 455-468.
[16] Barretti, P;Montelli, AC; Batalha, JEN; Caramori, JCT; Cunha, MLRS. The role of virulence factors in the outcome of staphylococcal peritonitis in CAPD patients. *BMC Infect Dis.*, 2009,9, 212-9.
[17] Haslinger-Löffler, B; Wagner, B; Brück, M; Strangfeld, K; Grundmeier, M; Fischer, U;et al. *Staphylococcus aureus* induces caspase-independent cell death in human peritoneal mesothelial cells. *Kidney Int.*, 2006, 70(6), 1089-98.
[18] Barretti, P; Moraes, TMC; Camargo, CH; Caramori, JCT; Mondelli, AL; Montelli, AC; Cunha, MLRS. Peritoneal dialysis-related peritonitis due to *Staphylococcus aureus*: a single-center experience over 15 years. *Plos One.*, 2012,7, e31780.
[19] Hayashida, A; Bartlett, AH; Foster, TJ; Park, PW. *Staphylococcusaureus* beta-toxin induces lung injury through syndecan-1. *Am J Pathol.*, 2009, 174(2), 509-18.
[20] Huseby, MJ; Kruse, AC; Digre, J; Kohler, PL; Vocke, JA; Mann, EE; et al. Beta toxin catalyzes formation of nucleoprotein matrix in staphylococcal biofilms. *Proc Natl Acad Sci U S A.*, 2010, 107(32), 14407-12.
[21] Boyle-Vavra, S; Daum, RS. Community-acquired methicillin-resistant *Staphylococcus aureus*: the role of Panton-Valentine leukocidin. *Lab Invest.*, 2007, 87(1), 3-9.
[22] Pimenta-Rodrigues, MV. Epidemiologia molecular e fatores de risco para aquisição de clones endêmicos de *Staphylococcus aureus* resistente à meticilina (MRSA) em um hospital de ensino. [Tese]. Botucatu: Universidade Estadual Paulista Júlio de Mesquita Filho: 2011.
[23] Bonesso, MF. Determinação da Virulência e da Resistência Antimicrobiana em *Staphylococcus* spp. Isolados de Pacientes do Serviço de Dermatologia do Hospital das Clínicas da Faculdade de Medicina de Botucatu, SP. [Dissertation]. Botucatu: Universidade Estadual Paulista Júlio de Mesquita Filho: 2011.

[24] Bergdoll, MS; Crass, BA; Reiser, RF; Robbins, RN; Davis, JP. A new staphylococcal enterotoxin, enterotoxin F, associated with toxic-shock syndrome *Staphylococcus aureus* isolates. *Lancet.*, 1981, 1, 1017-21.
[25] Bergdoll, MS; Borja, CR; Robbins, RN; Weiss, KF. Identification of enterotoxin E. *Infect Immun.*, 1971, 4(5), 593-5.
[26] Bergdoll, MS; Surgalla, MJ; Dack, GMA. Staphylococcal enterotoxin. Identification of a specific precipitating antibody with enterotoxin-neutralizing property. *J Immunol.*, 1959, 83, 334-8.
[27] Schlievert, PM; Case, LC. Molecular analysis of staphylococcal superantigens. *Methods Mol Biol.*, 2007, 391, 113-26.
[28] Larkin, EA; Carman, RJ; Krakauer, T; Stiles, BG. *Staphylococcus aureus*: the toxic presence of a pathogen extraordinaire. *Curr Med Chem.*, 2009, 16(30), 4003-19.
[29] Le, Loir, Y; Baron, F; Gautier, M. *Staphylococcus aureus* and food poisoning. *Genet Mol Res.*, 2003, 2(1), 63-76.
[30] Ono, HK; Omoe, K; Imanishi, K; Iwakabe, Y; Hu, DL; Kato, H; et al. Identification and characterization of two novel staphylococcal enterotoxins, types S and T. *Infect Immun.*, 2008, 76(11), 4999-5005.
[31] Lisa, RW. *Staphylococcus aureus* exfoliative toxins: How they cause disease. *Derm. Foundat.*, 2004,122, 1070-1077.
[32] Omoe, K; Imanishi, K; Hu, DL; Kato, H; Fugane, Y; Abe, Y; et al. Characterization of novel staphylococcal enterotoxin-like toxin type P. *Infect Immun.*, 2005, 73(9), 5540-6.
[33] Bergdoll, MS; Schlievert, PM. Toxic shock syndrome toxin. *Lancet.*, 1984,2, 69.
[34] Novick, RP. Mobile genetic elements and bacterial toxinoses: the superantigen-encoding pathogenicity islands of *Staphylococcus aureus*. *Plasmid.*, 2003, 49(2), 93-105.
[35] Lindsay, JA; Holden, MT. *Staphylococcus aureus*: superbug, super genome? *Trends Microbiol.*, 2004, 12(8), 378-85.
[36] Schmidt, H; Hensel, M. Pathogenicity islands in bacterial pathogenesis. *Clin Microbiol Rev.*, 2004, 17(1), 14-56.
[37] Lindsay, JA; Ruzin, A; Ross, HF; Kurepina, N; Novick, RP. The gene for toxic shock toxin is carried by a family of mobile pathogenicity islands in *Staphylococcus aureus*. *Mol Microbiol.*, 1998, 29(2), 527-43.
[38] Jarraud, S; Peyrat, MA; Lim, A; Tristan, A; Bes, M; Mougel, C; et al. *egc*, a highly prevalent operon of enterotoxin gene, forms a putative nursery of superantigens in *Staphylococcus aureus*. *J Immunol.*, 2001, 166(1), 669-77.

[39] Kuroda, M; Ohta, T; Uchiyama, I; Baba, T; Yuzawa, H; Kobayashi, I; et al. Whole genome sequencing of methicillin-resistant *Staphylococcus aureus*. *Lancet.*, 2001, 357(9264), 1225-40.

[40] Baba, T; Takeuchi, F; Kuroda, M; Yuzawa, H; Aoki, K; Oguchi, A; et al. Genome and virulence determinants of high virulence community-acquired MRSA. *Lancet.*, 2002, 359(9320), 1819-27.

[41] Becker, K; Friedrich, AW; Peters, G; vonEiff, C. Systematic survey on the prevalence of genes coding for staphylococcal enterotoxins SElM, SElO, and SElN. *Mol Nutr Food Res.*, 2004, 48(7), 488-95.

[42] Betley, MJ; Mekalanos, JJ. Staphylococcal enterotoxin A is encoded by phage. *Science.*, 1985, 229(4709), 185-7.

[43] Couch, JL; Soltis, MT; Betley, MJ. Cloning and nucleotide sequence of the type E staphylococcal enterotoxin gene. *J Bacteriol.*, 1988, 170(7), 2954-60.

[44] Bayles, KW; Iandolo, JJ. Genetic and molecular analyses of the gene encoding staphylococcal enterotoxin D. *J Bacteriol.*, 1989, 171(9), 4799-806.

[45] Zhang, S; Iandolo, JJ; Stewart, GC. The enterotoxin D plasmid of *Staphylococcus aureus* encodes a second enterotoxin determinant (*sej*). *FEMS Microbiol Lett.*, 1998, 168(2), 227-33.

[46] Omoe, K; Hu, DL; Takahashi-Omoe, H; Nakane, A; Shinagawa, K. Identification and characterization of a new staphylococcal enterotoxin-related putative toxin encoded by two kinds of plasmids. *Infect Immun.*, 2003, 71(10), 6088-94.

[47] van, Belkum, A:, Melles, DC; Snijders, SV:, vanLeeuwen, WB; Wertheim, HFL; Jan, L., Nouwen, JL:, Verbrugh, HA; Etienne, J. Clonal distribution and differential occurrence of the enterotoxin gene cluster, *egc*, in carriage- versus bacteremia-associated isolates of *Staphylococcus aureus*. *J. Clin. Microbiol.*, 2006, 44(4), 1555-1557.

[48] Bergdoll, MS. Analytical methods for *Staphylococcus aureus*. *Int J Food Microbiol.*, 1990, 10(2), 91-9.

[49] Kokan, NP; Bergdoll, MS. Detection of low-enterotoxin-producing *Staphylococcus aureus* strains. *Appl Environ Microbiol.*, 1987, 53(11), 2675-6.

[50] Park, CE; Szabo, R. Evaluation of the reversed passive latex agglutination (RPLA) test kits for detection of staphylococcal enterotoxins A, B, C, and D in foods. *Can J Microbiol.*, 1986, 32(9), 723-7.

[51] Cunha-Souza, MLR. Estafilococos enterotoxigênicos: efeito de cultura mista em leite, extrato de soja e parâmetros causadores de injuria celular. [Dissertation]. Londrina: Universidade Estadual de Londrina – UEL: 1992.
[52] Cremonesi, P; Luzzana, M; Brasca, M; Morandi, S; Lodi, R; Vimercati, C; et al. Development of a multiplex PCR assay for the identification of *Staphylococcus aureus* enterotoxigenic strains isolated from milk and dairy products. *Mol Cell Probes.*, 2005, 19(5), 299-305.
[53] Poli, MA; Rivera, VR; Neal, D. Sensitive and specific colorimetric ELISAs for *Staphylococcus aureus* enterotoxins A and B in urine and buffer. *Toxicon.*, 2002, 40(12), 1723-6.
[54] Sá, MEP;Cunha, MLRS; Elias, AO; Victória, C; Langoni, H; Paes, AC. Importance of *Staphylococcus aureus* in bovine subclinical mastitis: presence of enterotoxins, shock syndrome toxin and relationship with somatic cell count. *Braz J Vet Res Anim Sci.*, 2004, 41(5), 321-326.
[55] Johnson, WM; Tyler, SD; Ewan, EP; Ashton, FE; Pollard, DR; Rozee, KR. Detection of genes for enterotoxins, exfoliative toxins, and toxic shock syndrome toxin 1 in *Staphylococcus aureus* by the polymerase chain reaction. *J Clin Microbiol.*, 1991, 29(3), 426-30.
[56] Crass, BA; Bergdoll, MS. Involvement of coagulase-negative Staphylococci in toxic shock syndrome. *J Clin Microbiol.*, 1986, 23(1), 43-5.
[57] Kahler, RC; Boyce, JM; Bergdoll, MS; Lockwood, WR; Taylor, MR. Toxic shock syndrome associated with TSST-1 producing coagulase-negative Staphylococci. *Am J Med Sci.*, 1986, 292(5), 310-2.
[58] Valle, J; Gomez-Lucia, E; Piriz, S; Goyache, J; Orden, JA; Vadillo, S. Enterotoxin production by staphylococci isolated from healthy goats. *Appl Environ Microbiol.*, 1990, 56(5), 1323-6.
[59] Valle, J; Vadillo, S; Piriz, S; Gomez-Lucia, E. Toxic shock syndrome toxin 1 (TSST-1) production by Staphylococci isolated from goats and presence of specific antibodies to TSST-1 in serum and milk. *Appl Environ Microbiol.*, 1991, 57(3), 889-91.
[60] Rosec, JP; Gigaud, O. Staphylococcal enterotoxin genes of classical and new types detected by PCR in France. *Int J Food Microbiol.*, 2002, 77(1-2), 61-70.
[61] Cunha, MLRS; Peresi, E; Calsolari, RA; Araújo, Júnior, JP. Detection of enterotoxins genes in coagulase-negative staphylococci isolated from foods. *Braz J Microbiol.*, 2006,37(1),70-4.

[62] Udo, EE; Al-Bustan, MA; Jacob, LE; Chugh, TD. Enterotoxin production by coagulase-negative Staphylococci in restaurant workers from Kuwait City may be a potential cause of food poisoning. *J Med Microbiol.*, 1999, 48(9), 819-23.

[63] Marín, ME; de, la, Rosa, MC; Cornejo, I. Enterotoxigenicity of *Staphylococcus* strains isolated from Spanish dry-cured hams. *Appl Environ Microbiol.*, 1992, 58(3), 1067-9.

[64] Schmitz, FJ; Steiert, M; Hofmann, B; Verhoef, J; Hadding, U; Heinz, HP; et al. Development of a multiplex-PCR for direct detection of the genes for enterotoxin B and C, and toxic shock syndrome toxin-1 in *Staphylococcus aureus* isolates. *J Med Microbiol.*, 1998, 47(4), 335-40.

[65] Cunha, MLRS; Calsolari, RA; Araújo, Júnior, JP. Detection of enterotoxin and toxic shock syndrome toxin 1 genes in *Staphylococcus*, with emphasis on coagulase-negative staphylococci. *Microbiol Immunol.*, 2007, 51, 381-90.

[66] Becker, K; Haverkämper, G; vonEiff, C; Roth, R; Peters, G. Survey of staphylococcal enterotoxin genes, exfoliative toxin genes, and toxic shock syndrome toxin 1 gene in non-*Staphylococcus aureus* species. *Eur J Clin Microbiol Infect Dis.*, 2001, 20(6), 407-9.

[67] Kreiswirth, BN; Schlievert, PM; Novick, RP. Evaluation of coagulase-negative Staphylococci for ability to produce toxic shock syndrome toxin 1. *J Clin Microbiol.*, 1987, 25(10), 2028-9.

[68] Lotter, LP; Genigeorgis, CA. Deoxyribonucleic acid base composition and biochemical properties of certain coagulase-negative enterotoxigenic cocci. *Appl Microbiol.*, 1975, 29(2), 152-8.

[69] Fox, LK; Besser, TE; Jackson, SM. Evaluation of a coagulase-negative variant of *Staphylococcus aureus* as a cause of intramammary infections in a herd of dairy cattle. *J Am Vet Med Assoc.*, 1996, 209(6), 1143-6.

[70] Calsolari, RAO: Pereira, VCP; Araújo, Júnior, JP; Cunha, MLRS. Determination of toxigenic capacity by reverse transcription polymerase chain reaction in coagulase-negative staphylococci and *Staphylococcus aureus* isolated from newborns in Brazil. *Microbiol Immunol.*, 2011, 55, 394-407.

[71] Vasconcelos, NG; Pereira, VC; Araújo, Júnior, JP; Cunha, MLRS. Molecular detection of enterotoxins E, G, H and I in *Staphylococcus aureus* and coagulase-negative staphylococci isolated from clinical samples of newborns in Brazil. *J Appl Microbiol.*, 2011, 111, 749-62.

[72] Maiques, E; Ubeda, C; Tormo, MA; Ferrer, MD; Lasa, I; Novick, RP; et al. Role of staphylococcal phage and SaPI integrase in intra- and interspecies SaPI transfer. *J Bacteriol.*, 2007, 189(15), 5608-16.
[73] Madhusoodanan, J; Seo, KS; Remortel, B; Park, JY; Hwang, SY; Fox, LK; Park, YH; Deobald, CF; Wang, D; Liu, S; Daugherty, SC; Gill, AL; Bohach, GA; Gill, SR. An Enterotoxin-Bearing Pathogenicity Island in *Staphylococcus epidermidis*. *J Bacteriol.*, 2011.193(8), 1854–1862.
[74] Lindsay, JA. *Staphylococcus*: Molecular Genetics". Caister Academic Press. 1st ed. United Kingdom: 2008.
[75] Novick, RP; Jiang, D. The staphylococcal saeRS system coordinates environmental signals with *agr* quorum sensing. *Microbiology.*, 2003, 149(Pt 10), 2709-17.
[76] Novick, RP; Geisinger, E. Quorum sensing in Staphylococci. *Annu Rev Genet.*, 2008, 42, 541-64.
[77] Bohach, GA; Dinges, MM; Mitchell, DT; Ohlendorf, DH; Schlievert, PM. Exotoxins. In: Crosseley KB, Archer GL, editors. The *Staphylococci* in Human Disease. New York: Churchill Livingstone: 1997, 83–111.
[78] Ji, G; Beavis, R; Novick, RP. Bacterial interference caused by autoinducing peptide variants. *Science.*, 1997, 276(5321), 2027-30.
[79] McCuclloch, JA. Avaliação da funcionalidade do locus acessory gene regulator(*agr*) em cepas de *Staphylococcus aureus* brasileiras com suscetibilidade reduzida aos glicopeptídeos [Thesis]. São Paulo: Faculdade de Ciências Farmaceuticas, Universidade de São Paulo: 2006.
[80] Sung, JM; Chantler, PD; Lloyd, DH. Accessory gene regulator locus of *Staphylococcus intermedius*. *Infect Immun.*, 2006, 74(5), 2947-56
[81] Vandenesch, F; Projan, SJ; Kreiswirth, B; Etienne, J; Novick, RP. Agr-related sequences in *Staphylococcus lugdunensis*. *FEMS Microbiol Lett.*, 1993, 111(1), 115-22.
[82] van, Wamel, WJ; van, Rossum, G; Verhoef, J; Vandenbroucke-Grauls, CM; Fluit, AC. Cloning and characterization of an accessory gene regulator (*agr*)-like locus from *Staphylococcus epidermidis*. *FEMS Microbiol Lett.*, 1998,163(1), 1–9.
[83] Cheung, GYC; Joo, H-S; Chatterjee, SS; Otto, M. Phenol-soluble modulins – critical determinants of staphylococcal virulence. *FEMS Microbiol Rev.*, 2014,doi: 10.1111/1574-6976.12057.
[84] Queck, SY; Khan, BA; Wang, R; Bach, TH; Kretschmer, D; et al. Mobile genetic element-encoded cytolysin connects virulence to methicillin resistance in MRSA. *PLoS Pathog.*, 2009,5, e1000533.

[85] Wang, R; Braughton, KR; Kretschmer, D; Bach, TH; Queck, SY; et al. Identification of novel cytolytic peptides as key virulence determinants for community-associated MRSA. *Nat Med.*, 2007,13, 1510–4
[86] Wang, R; Khan, BA; Cheung, GYC; Bach, TL; Jameson-Lee, M; Kong, K; Queck, AY; Otto, M. *Staphylococcus epidermidis* surfactant peptides promote biofilm maturation and dissemination of biofilm-associated infection in mice. *J Clin Invest.*, 2011, 121(1), 238–248.
[87] Periasamy, S; Chatterjee, SS; Chatterjee, GYC; Otto, M. Phenol-soluble modulins in staphylococci. What are they originally for? *Commun Integr Biol.*, 2012, 5(3), 275-7.
[88] Schlievert, PM. Role of superantigens in human disease. *J Infect Dis.*, 1993, 167, 997-1002.

Chapter 7

ANTIMICROBIAL RESISTANCE

The acquisition of resistance to various antimicrobial agents has become a major problem in the treatment of nosocomial and community-acquired infections caused by *Staphylococcus* spp. In the pre-antibiotic era, the prognosis of severe staphylococcal infections was extremely poor. The introduction of penicillin as a therapeutic option in 1944 temporarily resolved the problem of these infections. However, the first resistant strains emerged in 1946, with 6% of *S. aureus* isolates producing penicillinase. In 1948, more than 50% of nosocomial *S. aureus* isolates were resistant to penicillin [1]. This proportion subsequently increased to about 80-90% [2]. The progressive dissemination of these strains drastically reduced the therapeutic value of this antibiotic and, to date, only a small percentage of *S. aureus* isolates continue to be susceptible. Similar data are observed in Brazil [3] and, according to a local study involving *S. aureus* strains isolated from blood cultures of patients seen at the University Hospital of the Botucatu Medical School (Hospital das Clínicas da Faculdade de Medicina de Botucatu – HC-FMB), 93% of *S. aureus* isolates are resistant to this drug [4].

Coagulase-negative staphylococci (CoNS) also exhibit high resistance to penicillin, with rates of about 70% reported in the study of Tavares [5] and > 70% reported in a study on 745 CoNS strain isolated from a university hospital in China between 2004 and 2009 [6]. Ustulin and Cunha [7] observed the production of beta-lactamase by 89.3% of CoNS strains isolated from blood cultures. Resistance to penicillin is attributed to the production of enzymes that inhibit the action of the drug. These enzymes, which hydrolyze the beta-lactam ring of penicillins, are known as penicillinases or, more generically, beta-lactamases [8]. The production of penicillinase is mediated by plasmids, but integration of this gene into the chromosome is common [9].

The introduction of methicillin and other semi-synthetic penicillins, such as penicillinase-resistant oxacillin and methicillin, in 1959 was a significant step in antistaphylococcal therapy. However, resistance to these drugs was detected about 2 years later [10]. The methicillin-resistant strains spread rapidly and their frequency has increased in different geographic regions, with these isolates being responsible particularly for outbreaks of nosocomial infections [9]. When resistance was described in 1961, methicillin was used in susceptibility tests and for the treatment of *S. aureus* infections. However, in the early 1990s, oxacillin, which belongs to the same class of drugs, was selected as the agent of choice for treatment and susceptibility tests. The acronym MRSA (methicillin-resistant *Staphylococcus aureus*) continues to be used to describe resistance due to its historical role [2].

Intrinsic resistance of *S. aureus* to oxacillin is mediated by the production of a supplemental penicillin-binding protein (PBP2' or PBP2a), which exhibits low affinity for semi-synthetic penicillins. The genetic determinant of this protein, the *mecA* gene, is located on the chromosome. To destroy bacteria, many antibiotics bind to PBPs and inactivate them. These proteins are involved in the formation of the bacterial cell wall. In the absence of a correctly assembled cell wall, bacteria are unable to maintain their integrity and die. Normally, staphylococci use three PBPs (1, 2 and 3) for cell wall synthesis, whereas methicillin- or oxacillin-resistant staphylococci (MRSA) possess a supplemental PBP, PBP2' or PBP2a. Therefore, when the *mecA* gene is present, the cell is able to grow in the presence of oxacillin and other beta-lactam antibiotics [9, 11].

The *mecA* gene is carried by a mobile genetic element identified as staphylococcal cassette chromosome *mec* (SCC*mec*). This element is integrated at a specific site in the chromosome, called *orfx*, and comprises the *mec* gene complex, the *ccr* gene complex and J regions. The *mec* gene complex consists of the *mecA* gene and its regulatory genes *mecI* and *mecRI*. The *ccr* gene complex is responsible for the integration and excision of SCC*mec* in the chromosome. In contrast, the J regions are not essential for the cassette chromosome, but can carry genes encoding resistance to non-beta-lactam antibiotics and heavy metals [12]. Eleven types of SCC*mec* have been described so far (Table 1), which are defined by the combination of the type of *ccr* gene complex and the class of *mec* gene complex. Subtypes are defined by polymorphisms in the J region of the same combination of *mec* and *ccr* complexes [12].

Table 1. Types of SCCmec identified in *Staphylococcus aureus*

SCCmec type	mec gene complex	ccr gene complex	Additional genetic elements
I	B(IS431-mecA-ΔmecR1-ψAS1272)	1 (A1B1)	
II	A(IS431-mecA-ψmecR1-mecI)	2 (A2B2)	pUB110, Tn554
III	A(IS431- mecA-ψmecR1-mecI)	3 (A3B3)	SCC$_{Hg}$, pI258, Tn554, pT181, ψTn554
IV	B(IS431 mecA-ΔmecR1-ψAS1272)	2 (A2B2)	
V	C2(IS431-mecA-ΔmecR1-IS431)*	5 (C1)	
VI	B(IS431-mecA-ΔmecR1-ψAS1272)	4 (A4B4)	
VII	C1(IS431-mecA-ΔmecR1-IS431)**	5 (C1)	
VIII	A(*IS431-mecA-ΔmecR1-*mecI)	4 (A4B4)	Tn554
IX	C2 (IS431-mecA-ΔmecR1-IS431)	1 (A1B1)	
X	C1(IS431-mecA-ΔmecR1-IS431)	7 (A1B6)	
XI	E(blaZ-mecA$_{LGA}$251-mecR1LGA251-mecILGA251)	8 (A1B3)	

*Two IS*431* arranged in the same direction.
**Two IS*431* arranged in opposite directions.
pUB110: carries the *ant(4')* gene which encodes resistance to different aminoglycosides (kanamycin, tobramycin, and bleomycin).
Tn554: carries the *ermA* gene which encodes constitutive and inducible resistance to macrolides, lincosamides and streptogramin (MLS).
SCC$_{Hg}$: staphylococcal cassette chromosome conferring resistant to heavy metals.
pI258: encodes resistance to penicillins and heavy metals (mercury).
pT181: encodes resistance to tetracycline.
ψTn554: encodes resistance to cadmium.

Although *mecA* gene-mediated resistance is present in all cells of a population with intrinsic resistance, it may only be expressed by a small proportion of these cells, a fact leading to the so-called heteroresistance. The expression of intrinsic resistance is classified into four classes, classes 1 to 4, where class 1 is the most heterogenous and class 4 the most homogenous [13].

In cultures of class 1 heteroresistant strains, most cells (99.9 to 99.99%) exhibit minimum inhibitory concentrations (MICs) of 1.5 to 3 μg/ml, but these cultures also contain a small number of bacteria (10^{-7} to 10^{-8}) that can form colonies even in the presence of 25 μg oxacillin/ml or higher. In cultures of class 2 strains, most cells ($\geq$ 99.9%) exhibit MICs of 6 to 12 μg/ml, and the frequency of highly resistant cells (those able to grow in the presence of 25 μg/ml) in these cultures is higher (10^{-5}) than in cultures of class 1 strains [13]. Cultures of class 3 strains are composed of bacteria (99 to 99.9%) exhibiting high levels of resistance to oxacillin (MIC = 50 to 200 μg/ml), but generally contain a subpopulation (10^{-3}) of highly resistant cells that are able to form colonies even in the presence of 300 to 400 μg oxacillin/ml. Cultures of class 4 strains consist of cells with a homogenous resistance pattern, in which all cells exhibit high levels of resistance to oxacillin (MIC of 400 to 1000 μg/ml) [13].

Another resistance modality has also been described in isolates that do not carry the *mecA* gene, which is called borderline resistance. Borderline resistance can be attributed to two mechanisms: the first consists of the inactivation of oxacillin mediated by the overproduction of beta-lactamase [14], and the second is modified resistance, called MOD-SA, which is due to the altered affinity of intrinsic PBPs for oxacillin [13]. These resistance modalities are characterized by a low level of resistance (MIC of 8 μg/ml) [11].

The phenotypic expression of *mecA* gene-mediated resistance is affected by different factors, such as pH, temperature and osmolarity [15]. Laboratory detection of MRSA is obtained without major difficulties when appropriate conditions are used, including the supplementation of Mueller-Hinton agar with NaCl and adequate temperatures and incubation times as recommended by the Clinical and Laboratory Standards Institute (CLSI) [16]. However, detection can be more difficult in the case of more heterogenous strains, even when reference methods are used [11].

The adequate detection of *mecA* gene-mediated oxacillin resistance is important for the clinical laboratory. Although recommended methods detect most oxacillin-resistance strains, there are two situations that require additional steps to confirm this susceptibility or resistance. The first situation is the occurrence of extremely heterogenous strains which are classified as susceptible by reference methods. The second is the occurrence of borderline resistance (MIC close to the susceptibility breakpoint), which should be distinguished from *mecA* gene-mediated resistance since the clinical significance of the latter is much greater. Experimental animal studies and clinical data have shown that beta-lactam antibiotics are effective in treating infections caused by isolates that do not carry the *mecA* gene and exhibit low levels of resistance (borderline) [17, 18]. However, infections caused by isolates carrying the *mecA* gene require treatment with vancomycin [19].

The reference methods recommended by the CLSI for detecting oxacillin resistance in *S. aureus* include the determination of MICs by the agar or broth dilution method, disk diffusion method, screening on Mueller-Hinton agar supplemented with 4% NaCl and 6 µg oxacillin, and cefoxitin disk diffusion test [16, 20]. The sensitivity of dilutions methods in detecting isolates with *mecA* gene-mediated resistance ranges from 98 to 100% [21]. Although few studies report the degree of heterogeneity of the isolates tested, it is supposed that resistant strains not detected by the dilution methods are more heterogenous [11].

Studies evaluating the performance of disk diffusion for detection of MRSA have shown that this method is less reliable in the case of heterogenous strains. In a study using confirmed heterogenous strains, the sensitivity of this method was 61% for a total of 80 *mecA*-positive isolates. Felten et al. [22] reported a sensitivity of 88.5% for class 1 strains (extremely heterogenous) and of 96.4% for class 2 strains [22]. Velasco et al. [23] showed that disk diffusion, E-test and microdilution methods were often not fully reliable in detecting isolates carrying the *mecA* gene. Three of 51 *mecA*-positive strains tested false negative by these methods. According to these authors, the cefoxitin disk (30 µg) as a screening method provided the best results, with 100% sensitivity and 98% specificity. Similar results have recently been reported by our group for *S. aureus* isolated from patients of the Pediatric Units of FMB, Brazil. Detection by the cefoxitin disk diffusion method showed 100% sensitivity and 98% specificity versus 94.4% and 98.8% for the oxacillin disk [24]. However, a study conducted by the same group using *S. aureus* strains isolated from blood cultures of adult patients seen at the same hospital demonstrated lower sensitivities for the disk diffusion methods (86.9% and 91.3% for the oxacillin and cefoxitin disk, respectively), and the same specificity (91.3%). The sensitivity of the screening test (91.3%) was similar to that of the cefoxitin disk and specificity was the same. The best results were obtained with the E-test, showing a sensitivity of 97.8% and the same specificity as the other methods [4].

Ustulin and Cunha [7], studying CoNS isolated from adult patients seen at HC-FMB, Brazil, reported an oxacillin resistance rate of 82.5%. The sensitivity obtained for the oxacillin and cefoxitin disk diffusion tests was the same (95.3%). The E-test showed the highest sensitivity (98.8%) and the same specificity (83.3%) as observed for the other methods. A study conducted at the same hospital, but testing blood cultures obtained from newborns hospitalized in the neonatal intensive care unit between 1991 and 2009, revealed a percentage of *mecA*-positive CoNS of 69%. The sensitivity of the oxacillin and cefoxitin disk diffusion tests was 92.8% and 91.3%, respectively, and specificity was 100% for the CoNS isolates. The sensitivity and specificity of the screening method were 92.8% and 100%, respectively [25].

Since phenotypic methods for the detection of MRSA may sometimes provide questionable results, molecular techniques for detecting the *mecA* gene or its product PBP 2a have been proposed. Investigation of the *mecA* gene by the polymerase chain reaction (PCR) is considered to be the gold standard for MRSA detection [20]. According to the CLSI [20], detection of the *mecA* gene or of the protein encoded by this gene (PBP2a) is the most adequate method to determine oxacillin resistance and can be used to confirm the results of disk tests in the case of more severe infections.

MRSA strains are usually resistance to other beta-lactam antibiotics, macrolides, aminoglycosides, chloramphenicol, quinolones, and tetracycline [26, 27]. On the basis of *in vitro* observations, the CLSI [20] recommends oxacillin-resistant *S. aureus* to be considered resistant to all beta-lactams, including cephalosporins and carbapenems. Infections caused by MRSA are thus treated with a glycopeptide antibiotic, vancomycin, which was introduced in 1968 and continues to be effective in the treatment of these infections.

With the development of vancomycin-resistant *Enterococcus* (VRE) in 1988 and the possibility of transfer of this resistance to *S. aureus*, the surveillance of vancomycin resistance has become a subject of scientific interest worldwide. In 1996, the first clinical isolate of *S. aureus* with reduced vancomycin susceptibility, with MICs in the intermediate range (8 µg/ml), called vancomycin-intermediate *S. aureus* (VISA), was reported in Japan [28]. Subsequently, in June 2002, eight patients with infections caused by *S. aureus* with reduced vancomycin susceptibility were confirmed in the United States [29]. One month later, the CDC reported the first case of vancomycin-resistant *S. aureus* (VRSA, MIC = ≥ 32 µg/ml) in a patient from Michigan, United States. The strain isolated from the patient carried the *vanA* gene and also the *mecA* gene encoding oxacillin resistance. The presence of the *vanA* gene in this VRSA isolate suggests that resistance may have been acquired by the transfer of genetic material from VRE to *S. aureus*. In October of the same year [29], the second clinical VRSA isolate was documented in a patient from Pennsylvania. This VRSA isolate also carried the *vanA* and *mecA* genes. The presence of the *vanA* gene again suggests that the resistant determinant was acquired from VRE which were isolated from the same patient. The third case of VRSA isolated from a patient in New York was reported in April 2004. The isolate also carried the *mecA* and *vanA* genes encoding oxacillin and vancomycin resistance, respectively. According to the CDC, the three VRSA isolates were epidemiologically unrelated [29, 30].

Table 2. History of cases of vancomycin-resistant *Staphylococcus aureus* in the United States and geographic information

Case	State	Year	Age (years)	Origin	Diagnosis	Underlying disease
1	Michigan	2002	40	Plantar ulcer and catheter	Soft tissue infection	Diabetes, dialysis
2	Pennsylvania	2002	70	Plantar ulcer	Osteomyelitis	Obesity
3	New York	2004	63	Nephrostomy tube	No infection	Multiple sclerosis, diabetes, kidney stones
4	Michigan	2005	78	Toe wound	Gangrene	Diabetes, vascular disease
5	Michigan	2005	58	Surgical site wound after panniculectomy	Surgical site infection	Obesity
6	Michigan	2005	48	Plantar ulcer	Osteomyelitis	Chronic ulcer
7	Michigan	2006	43	Triceps wound	Necrotizing fascitis	Diabetes, dialysis, chronic ulcer
8	Michigan	2007	48	Toe wound	Osteomyelitis	Diabetes, obesity, chronic ulcer
9	Michigan	2007	54	Surgical site wound after food amputation	Osteomyelitis	Diabetes, hepatic encephalopathy
10	Michigan	2009	53	Plantar wound	Soft tissue infection	Diabetes, obesity, lupus, rheumatoid arthritis
11	Delaware	2010	64	Wound drainage	Prosthetic infection	Diabetes, kidney disease, dialysis
12	Delaware	2010	83	Vaginal swab	Vaginal discharge	Chronic infection with *C. difficile*, chronic urinary tract infection
13	Delaware	2012	70	Foot wound	Chronic wound, possible osteomyelitis	Outpatient with a chronic wound, hypertension and diabetes mellitus

Source: CDC [31, 32].

The CDC [31, 32] recently confirmed the 13th case of VRSA in the United States (Table 2). These data highlight the important role of clinical laboratories in the diagnosis of cases of VRSA to ensure the prompt recognition, isolation and control of infection.

The prescription of the appropriate antibiotic by health professionals and the compliance with guidelines for the control and surveillance of MRSA and VRE are necessary to prevent the further emergence of VRSA strains.

In Brazil, VISA strains have been described by Oliveira et al. [33]. Four strains were isolated from patients of a Burn Unit and one strain from a patient of an Orthopedics Unit, all of them treated with vancomycin for more than 30 days. The development of intermediate susceptibility to vancomycin might be related to prolonged contact of the microorganism with this antibiotic, a fact raising epidemiological concern regarding the detection and control of this resistance in Brazilian hospitals. Intermediate vancomycin resistance may involve metabolic alterations, including thickening of the bacterial cell wall due to the accelerated synthesis of peptidoglycans. As a consequence, additional D-alanyl-D-alanine sites are available for the binding of vancomycin, which will eventually be depleted and no longer inhibits peptidoglycan synthesis, leading to thickening of the cell wall [34]. Recently, the first VRSA was reported in Brazil, which was isolated from a patient with bloodstream infection caused by an MRSA strain that was susceptible to vancomycin, but that acquired the *vanA* gene cluster during antibiotic therapy and became resistant to vancomycin [35].

As observed for *S. aureus*, when CoNS become resistant to the antibiotics most commonly used, the glycopeptide vancomycin is the drug of choice [36]. However, Veach et al. [37] isolated vancomycin-resistant *S. haemolyticus* strains from patients undergoing prolonged therapy with this antibiotic. The *S. haemolyticus* strains isolated from the patients under treatment with vancomycin exhibited a reduction in the susceptibility to this drug when compared to strains isolated before antibiotic therapy [37]. Although rare, these isolates may be a sign of the beginning of resistance to an important antibiotic used to treat staphylococcal infections.

In 2006, a vancomycin-resistant *S. cohnii* strain was isolated from the pleural fluid of a 5-year-old child treated at the San Jerônimo de Monteria Hospital in Colombia. This patient had been under prolonged treatment with vancomycin and the MIC determined by the E-test was 64 µg/ml. The same isolate was resistant to oxacillin, teicoplanin, ceftazidime, and trimethoprim-sulfamethoxazole [38].

Another report, this time involving CoNS strains isolated from healthy individuals, was published by Palazzo et al. [39] in Brazil. Four vancomycin-resistant *Staphylococcus* spp. (1 *S. epidermidis*, 1 *S. haemolyticus*, and 2 *S. capitis*) were isolated from employees of a private school and from a hospital located in the region of Ribeirão Preto, state of São Paulo. The MIC of these isolates determined by the E-test ranged from 16 to ≥ 256 μg/ml. In that study, the *van* genes were not responsible for resistance, since neither the *vanA* nor the *vanB* gene could be detected, suggesting that bacterial cell wall thickening may have played a role in vancomycin resistance in these isolates.

The choice of an adequate laboratory method influences the ability of glycopeptide resistance surveillance and explains why the effective global prevalence of isolates is still unknown [2]. Although the disk diffusion test is the most common method used in routine laboratory testing, it is not indicated for the detection of VISA or VRSA [20, 31]. The methods recommended by the CDC [31] and CLSI [20] are the determination of MICs or screening on BHI agar containing 6 μg vancomycin/ml. All *Staphylococcus* spp. that grow in the screening test should be confirmed in relation to the presence of pure culture and tested by a method for the determination of vancomycin MIC. As observed for MRSA, VISA isolates express heterogenous resistance. Colonies that express vancomycin resistance grow slowly, a fact impairing detection by the routine methods used for susceptibility testing in most clinical laboratories (disk diffusion). *Staphylococcus aureus* with MICs of 2 to 4 μg/ml should therefore be analyzed carefully [2]. The treatment options for *Staphylococcus* spp. with reduced vancomycin susceptibility are limited and infection control precautions are needed to reduce transmission and to minimize possible outbreaks.

Other more recent classes of antibiotics, such as streptogramins (quinupristin/dalfopristin), oxazolidines (linezolide), glycylcyclines (tigecycline) and lipopeptides (daptomycin), are options for the treatment of VISA or VRSA infections. However, some specific mechanisms of resistance to these new antibiotics have already been selected in staphylococci. Although these isolates are rare, from a pessimist's point of view one may say that, since resistant strains exist, the useful life of these new antibiotics is already threatened. From an optimist's point of view, the availability of several antimicrobial alternatives against multidrug-resistant microorganisms may delay the dissemination of resistance [40]. However, the careful use of these new classes is imperative to preserve these therapeutic options.

REFERENCES

[1] Barber, M; Rozwadowska-Dowzenko, M. Infection by penicillin-resistant staphylococci. *Lancet.* 1948, 2(6530), 641-4.

[2] CDC. Centers of Diseases Control and Prevention. Laboratory capacity to detect antimicrobial resistance. *MMWR.* 2000, 48, 1167-71.

[3] Oliveira, GA; Faria, JB; Levy, CE; Mamizuka, EM. Characterization of the Brazilian endemic clone of methicillin-resistant *Staphylococcus aureus* (MRSA) from hospitals throughout Brazil. *Braz J Infect Dis.*, 2001, 5(4), 163-70.

[4] Martins, A; Pereira, VC; Cunha, MLRS. Oxacillin resistance of *Staphylococcus aureus* isolated from the University Hospital of Botucatu Medical School in Brazil. *Chemotherapy.*, 2010, 56, 112-9.

[5] Tavares, W. Problems with gram-positive bacteria: resistance in staphylococci, enterococci, and pneumococci to antimicrobial drugs. *Rev Soc Bras Med Trop.*, 2000, 33(3), 281-301.

[6] Ma, XX; Wang, EH; Liu, Y; Luo, EJ. Antibiotic susceptibility of coagulase-negative staphylococci (CoNS), emergence of teicoplanin-non-susceptible CoNS strains with inducible resistance to vancomycin. *J Med Microbiol.*, 2011, 60(Pt 11), 1661-8.

[7] Ustulin, DR; Cunha, MLRS. Methods for detection oxacillin resistance among coagulase-negative staphylococci recovered from patients with bloodstream infections at the University Hospital in Brazil. *J Virol Microbiol.*, 2012, 2012(2012), 1-12.

[8] Kloos; WE; Bannerman, TL. *Staphylococcus* and *Micrococcus*. In: Murray PR, Baron EJ, Pfaller, MA et al., editors. Manual of clinical microbiology. Washington: American Society for Microbiology, 1995. p. 282-298.

[9] Livermore, DM. Antibiotic resistance in staphylococci. *Int J Antimicrob Agents.*, 2000, 16 Suppl 1, S3-10.

[10] Hartman, BJ; Tomasz, A. Low-affinity penicillin-binding protein associated with beta-lactam resistance in *Staphylococcus aureus*. *J Bacteriol.*, 1984, 158(2), 513-6.

[11] Swenson, JM. New tests for the detection of oxacillin-resistant *Staphylococcus aureus*. *Clin. Microbiol. Newsletter.*, 2002, 24(21), 159-163.

[12] (IWG-SCC) International Working Group on the Classification of Staphylococcal Cassette Chromosome Elements. Classification of staphylococcal cassette chromosome *mec* (SCC*mec*), guidelines for

reporting novel SCC*mec* elements. *Antimicrob Agents Chemother.*, 2009, 53(12), 4961-7.
[13] Tomasz, A; Nachman, S; Leaf, H. Stable classes of phenotypic expression in methicillin-resistant clinical isolates of staphylococci. *Antimicrob Agents Chemother.*, 1991, 35(1), 124-9.
[14] McDougal, LK; Thornsberry, C. The role of the β-lactamases in staphylococcal resistance to penicillinase-resistant penicillins and cephalosporins. *J Clin Microbiol.*, 1986, 23, 832-9.
[15] Marty, V; Madiraju, VS; Brunner, DP; Wilkinson, BJ. Effects of temperature, NaCl, and methicillin on penicillin-binding proteins, growth, peptidoglycan synthesis, and autolysis in methicillin-resistant *Staphylococcus aureus*. Antimicrob. *Agents Chemother.*, 1987, 31, 727-33.
[16] CLSI. Clinical and Laboratory Standards Institute Performance standards for antimicrobial susceptibility testing. M100-S15., Wayne, PA., 2005.
[17] Massanari, RM; Pfaller, MA; Wakefield, DS; Hammons, GT; McNutt, LA; Woolson, RF; et al. Implications of acquired oxacillin resistance in the management and control of *Staphylococcus aureus* infections. *J Infect Dis.*, 1988, 158(4), 702-9.
[18] Thauvin-Eliopoulos, C; Rice, LB; Eliopoulos, GM; Moellering, RC. Efficacy of oxacillin and ampicillin-sulbactam combination in experimental endocarditis caused by beta-lactamase-hyperproducing *Staphylococcus aureus*. *AntimicrobAgents Chemother.*, 1990, 34(5), 728-32.
[19] Kolbert, CP; Connolly, JE; Lee, MJ; Persing, DH. Detection of the Staphylococcal *mecA* gene by chemiluminescent DNA hybridization. *J Clin Microbiol.*, 1995, 33(8), 2179-82.
[20] CLSI. Clinical and Laboratory Standards Institute. Performance Standards for Antimicrobial Susceptibility Testing, Twenty-First *Informational Supplement CLSI document.*, 2011,(M100-S21).
[21] Sakoulas, G; Gold, HS; Venkataraman, L; DeGirolami, PC; Eliopoulos, GM; Qian, Q. Methicillin-resistant *Staphylococcus aureus*: comparison of susceptibility testing methods and analysis of *mecA*-positive susceptible strains. *J Clin Microbiol.*, 2001, 39(11), 3946-51.
[22] Felten, A; Grandry, B; Lagrange, PH; Casin, I. Evaluation of three techniques for detection of low-level methicillin-resistant *Staphylococcus aureus* (MRSA), a disk diffusion method with cefoxitin

and moxalactam, the Vitek 2 system, and the MRSA-screen latex agglutination test. *J Clin Microbiol.*, 2002, 40(8), 2766-71.

[23] Velasco, D; del Mar Tomas, M; Cartelle, M; Beceiro, A; Perez, A; Molina, F; et al. Evaluation of different methods for detecting methicillin (oxacillin) resistance in *Staphylococcus aureus*. *J Antimicrob Chemother.*, 2005, 55(3), 379-82.

[24] Pereira, VC; Martins, A; Rugolo, LMSS; Cunha, MLRS. Detection of Oxacillin Resistance in *Staphylococcus aureus* isolated from the neonatal and Pediatric Units of a Brazilian Teaching Hospital. *Clin Med Pediatr.*, 2009, 3: 23-31.

[25] Pereira, VC; Cunha, MLRS. Coagulase-negative staphylococci strains resistant to oxacillin isolated from neonatal blood cultures. *Mem Inst Oswaldo Cruz.*, 2013, 108(7), 939-42.

[26] Turnidge, J; Grayson, ML. Optimum treatment of staphylococcal infections. *Drugs.*, 1993, 45(3), 353-66.

[27] Chambers, HF. Methicillin resistance in staphylococci: molecular and biochemical basis and clinical implications. *Clin Microbiol Rev.*, 1997, 10(4), 781-91.

[28] Hiramatsu, K; Hanaki, H; Ino, T; Yabuta, K; Oguri, T; Tenover, FC. Methicillin-resistant *Staphylococcus aureus* clinical strain with reduced vancomycin susceptibility. *J Antimicrob Chemother.*, 1997, 40(1), 135-6.

[29] CDC. Centers of Diseases Control and Prevention. *Staphylococcus aureus* resistant to vancomycin--United States, 2002. *MMWR MorbMortal Wkly Rep.*, 2002, 51(26), 565-7.

[30] CDC. Centers for Disease Control and Prevention. *Staphylococcus aureus* resistant to vancomycin-United States. *MMWR*, 2002, 51, 565-567. Brief Report: Vancomycin-Resistant *Staphylococcus aureus*- New York. *MMWR.*, 2004, 53, 322-323.

[31] CDC. Centers for Disease Control and Prevention. Reminds Clinical Laboratories and Healthcare Infection Preventionists of their Role in the Search and Containment of Vancomycin-Resistant *Staphylococcus aureus* (VRSA), May 2010. [Internet]. Accessed June 18, 2011 Available at: http://www.cdc.gov/HAI/settings/lab/vrsa_lab_search_containment.html.

[32] Limbago, BM; Kallen, AJ; Zhu, W; Eggers, P; McDougal, LK; Albrecht, VS. Report of the 13th Vancomycin-Resistant *Staphylococcus aureus* Isolate from the United States. *J. Clin. Microbiol.*, 2014, 52(3), 998-1002.

[33] Oliveira, GA; Levy, CE; Mamizuka, EM. Estudo do perfil de resistência de 626 cepas de *Staphylococcus aureus* isoladas de 25 hospitais brasileiros entre setembro de 1995 e junho de 1997. *J Bras Patol.*, 2000, 36, 147-55.

[34] Appelbaum, PC. The emergence of vancomycin-intermediate and vancomycin-resistant *Staphylococcus aureus*. *Clin Microbiol Infect.*, 2006, 12 Suppl 1, 16-23.

[35] Rossi, F; Diaz, L; Wollam, A; Panesso, D; Zhou, Y; Rincon, S; Narechania, A; Xing, G; Di, Gioia, TSR; Doi, A; Tran, TT; Reyes, J; Munita, JM; Carvajal, LP; Hernandez-Roldan, A; Brandão, D; van, der, Heijden, IM; Murray, BE; Planet, PJ; Weinstock, GM; Arias, CA. Transferable vancomycin resistance in a community-associated MRSA lineage. *N Engl J Med.*, 2014, 370, 1524-31.

[36] Kloos, WE; Bannerman, TL. Update on clinical significance of coagulase-negative Staphylococci. *Clin Microbiol Rev.*, 1994, 7(1), 117-40.

[37] Veach, LA; Pfaller, MA; Barrett, M; Koontz, FP; Wenzel, RP. Vancomycin resistance in *Staphylococcus haemolyticus* causing colonization and bloodstream infection. *J Clin Microbiol.*, 1990, 28(9), 2064-8.

[38] Martínez, P; Mattar, S. Posible aislamiento clínico de *Staphylococcus cohnii* resistente a vancomicina. *Infectio.*, 2006, 10, 175-177.

[39] Palazzo, IC; Araujo, ML; Darini, AL. First report of vancomycin-resistant staphylococci isolated from healthy carriers in Brazil. *J Clin Microbiol.*, 2005, 43(1), 179-85.

[40] Leclercq, R. Epidemiological and resistance issues in multidrug-resistant staphylococci and enterococci. *Clin Microbiol Infect.*, 2009, 15(3), 224-31.

Chapter 8

EPIDEMIOLOGY OF METHICILLIN-RESISTANT *STAPHYLOCOCCUS* SPP.

Endemic methicillin-resistant *Staphylococcus aureus* (MRSA) strains carrying multiple resistance determinants became a global problem in the early 1980s [1]. Recent scientific advances considering the genetic origin of methicillin resistance in *S. aureus* have boosted knowledge of the epidemiology of MRSA. The understanding of the epidemiology of MRSA infections has important implications for control measures. It is therefore necessary to document the dissemination of clones and to identify individual factors related to their acquisition. Molecular methods have been used to study the epidemiology of nosocomial MRSA infections on a local, national and international level. Molecular typing of microorganisms permits the characterization of microbial isolates at a molecular level, identifying any similarities between strains and characterizing their common evolutionary origin. A clone is defined as a set of genetically related bacterial isolates that are indistinguishable from each other by molecular typing methods, or isolates that are so similar that they are presumed to be derived from a common ancestor [2].

Using a combination of different molecular typing techniques, six MRSA clones spread worldwide have been identified. These clones, called Iberian, Brazilian, Hungarian, New York/Japan, pediatric and EMRSA-16, are responsible for 68% of MRSA infections and are successful isolates in terms of the ability to cause infection, persistence, and the capacity to spread from one geographic area to another and even between continents [3]. These clones were named according to the geographic area where they were first described (Brazilian epidemic clone, Iberian, New York/Japan), the epidemiological characteristics of the patients from which they were isolated (pediatric clone), the number based on their pulsed-field gel electrophoresis pattern (USA 100, USA 800), and the phage type (EMRSA-15, EMRSA-16) [4].

The Brazilian epidemic clone (BEC) was first described in 1993 [4] in a study conducted in the city of São Paulo and is found worldwide, including countries in Latin America and Europe [5, 6]. In Brazil, this clone has been identified in different hospitals throughout the country and is frequently associated with epidemic outbreaks [7, 8].

Molecular epidemiological studies highlight the continuing global evolution/spread of MRSA clones characterized by increasing antibiotic resistance and virulence. We only have a partial understanding of the factors that contribute to the spread of MRSA clones, but they are likely to include the migration of human populations, ineffective methods for the control of MRSA transmission by infected patients, and poorly effective treatment strategies including the inadequate use/choice of antibiotics. In hospitals, patients already infected with MRSA at the time of admission are more likely to develop infection caused by the colonizing bacteria or to transmit MRSA to other patients[9].

The rates of colonization or infection with MRSA vary according to geographic location, type of hospital service, and specific population. Colonized patients contaminate the environment with their strains, facilitating cross-transmission. This fact highlights the importance of precaution measures and the isolation of patients who are colonized or infected with MRSA, complemented by effective protocols for hygiene, cleaning and maintenance of hospital environments [10].

MRSA infections are associated with considerable morbidity and mortality. Additionally, they are more costly to manage than other infections [11]. The substantially higher expenses related to the management of these infections are due to prolonged hospitalization and increased care in isolation, as well as additional healthcare and the economic burden of secondary treatment. Some studies suggest the screening of high-risk patients for colonization with MRSA to be a cost-effective measure in order to limit the dissemination of these microorganisms in hospitals [12]. High success rates in the control of MRSA have been obtained in countries that adhered to strict policies of hospital infection control, including transmission-based precautions and restrictions to the use of antibiotics. As a consequence, the rapid and reliable identification of patients carrying MRSA is crucial for the strategy of infection control. Wernitz et al. [13] showed that, despite its cost, extensive patient screening for MRSA on admission to the hospital had an important positive impact on reducing MRSA infection rates.

The dissemination of MRSA in healthcare centers is difficult to control. In this respect, several international guidelines recommend measures that include active surveillance cultures for the identification of patients colonized or infected with MRSA, strict precaution and isolation measures, and quarantine of newly admitted patients until the possibility of MRSA colonization is ruled out [14]. The Netherlands is an example of a country where a strict "search and destroy" policy of MRSA is applied and where the endemic level of these microorganisms is low. All patients and health professionals are considered to be colonized until proven otherwise and are subjected to intensive and costly procedures of infection control. In this case, rapid MRSA screening is important to identify non-carriers in whom the rules regarding these procedures can be relaxed.

Staphylococcal cassette chromosome *mec* (SCC*mec*) typing is a useful epidemiological tool [15], since different types are more prevalent in the hospital environment or community. Community-associated MRSA (CA-MRSA) are potentially emerging pathogens that have shown an increasing frequency of isolates [16]. Patients infected with CA-MRSA are characterized by the fact that they have neither been hospitalized in the year prior to infection nor have been submitted to medical procedures such as dialysis, surgery or catheter insertion, events that are common in healthcare-associated MRSA (HA-MRSA) infection. Whereas HA-MRSA isolates are characterized by broad resistance to different antibiotics, CA-MRSA isolates are susceptible (85 to 100%) to drugs such as clindamycin, gentamicin, ciprofloxacin, sulfamethoxazole/trimethoprim and vancomycin, and are resistant only to oxacillin and other beta-lactams [3].

The difference between the resistance profiles of HA-MRSA and CA-MRSA isolates seems to be explained by the size and distribution of SCC*mec* types that carry the oxacillin resistance determinant. Among the main SCC*mec* types (I, II, III, IV and V), only types I, II and III are found in HA-MRSA isolates, whereas types IV and V are observed in CA-MRSA. Type IV has a smaller size and lower metabolic cost, characteristics that render this element selectively favored for transfer between *Staphylococcus* strains [17].

Although CA-MRSA generally cause skin infections, invasive diseases such as bacteremia, endocarditis, osteomyelitis and pneumonia have already been described, as well as outbreaks of nosocomial infections [18, 19]. Panton-Valentine leukocidin (PVL)-positive CA-MRSA are easily transmitted among families, and also on a large scale in the community, such as prisons, schools and sport teams. Skin-skin contact involving abrasions and indirect contact with contaminated objects such as towels, linen and sports equipment seem to be modes of transmission [3].

It is possible that some MRSA clones are more likely to cause disease than others. as a result of the presence of virulence factors that increase their chances to reach normally sterile sites, and to survive, proliferate and disseminate in the host. These factors include the secretion of exotoxins, hemolysins and leukocidins, as well as the production of biofilms [20]. Although many virulence factors have been identified in the genome of *S. aureus*, the differences in the pathogenic and invasive potential of isolates disseminated in the environment are still unknown. Our group recently compared the presence of genes encoding virulence factors and methicillin resistance between *S. aureus* isolates recovered from surveillance (colonizers) and clinical (invasive) cultures of patients seen at a small teaching hospital. The presence of SCC*mec* types III and IV, as well as of genes encoding exfoliative toxin B and PVL, was independently associated with invasion [21]. The detection of these genes in *S. aureus* isolates from colonized patients may be used as an indicator of those who need closer follow-up and intensive measures of infection control.

Pulsed-field gel electrophoresis (PFGE) continues to be the most commonly used method in micro-epidemiological studies (local outbreaks). The method is based on the direct digestion of genomic DNA with a restriction enzyme (typically *SmaI*), followed by separation of the fragments by gel electrophoresis with alternating electrical fields. The results obtained by our group in a study [22] of burn patients of a university hospital to determine the clonal and virulence profile of MRSA isolates by PFGE, and of a case-case-control study to evaluate risk factors for the acquisition of MRSA and MSSA revealed a polyclonal profile of MRSA in the burn unit, possibly related to imported strains. Most of the isolates (94.8%) harbored SCC*mec* type III and surgical procedures were indicated as risk factors for the dissemination of MRSA [22]. These results agree with another study [23] conducted by our group at the University Hospital of the Botucatu Medical School (Hospital das Clínicas da Faculdade de Medicina de Botucatu – HC-FMB), Brazil, with the most of the SCC*mec* III MRSA (89.1%). Molecular typing identified four MRSA clones. The major clone showed similarity > 80% with the BEC (BEC-HU 25), demonstrating that this clone, which is still prevalent in nosocomial infections in Brazil, is also present in our hospital [23].

A study conducted at São Paulo Hospital, which belongs to the Federal University of São Paulo (Universidade Federal de São Paulo – UNIFESP), analyzed isolates recovered from 2002 to 2005 and observed persistence of the BEC within the hospital environment, in addition to the emergence of new clones in the last years studied that differ from BEC, indicating temporal evolution of this clone in nosocomial environments [24]. Similar results have been reported in a study conducted in the Pediatric Unit of HC-FMB, Brazil, in which most MRSA isolates carried SCC*mec* type III (60%) and were related to BEC [25]. However, an increase in MRSA carrying SCC*mec* type IV and a reduction in SCC*mec* type III isolates have been observed over the years [25], confirming the global trend towards an increase in the frequency of SCC*mec* type IV in the hospital environment and a reduction in the prevalence of hospital clones [26, 27].

In view of their complex epidemiology, the presence of CA-MRSA in hospitals and the circulation of HA-MRSA isolates in the community, a clear delimitation between CA-MRSA and HA-MRSA is not possible. Researchers of the CDC have proposed a third category, healthcare-associated community-onset (HACO)-MRSA. This category includes cases that would be HA-MRSA based on the history of exposure to healthcare services, but acquisition starts in the community. This classification scheme (HA-, CA- and HACO-MRSA) still has limitations because a history of exposure to a healthcare environment does not exclude the possibility of acquisition of MRSA in the community [28, 29]. Therefore, revision of the nomenclature is needed to better reflect the contemporary epidemiology of MRSA.

A study conducted at our laboratory that involved patients with skin infections attending the Dermatology Clinic of HC-FMB, Brazil [30], identified *S. aureus* as the main causative agent of skin infections in patients from the community. MRSA accounted for 10.6% of all infections, with most of the isolates carrying SCC*mec* types IV and II. Previous studies reported these SCC*mec* types to be the most frequent types colonizing individuals in the community, with a predominance of type IV [31]. Despite the lower frequency of isolation from skin infections, coagulase-negative staphylococci (CoNS) presented a larger number of resistant strains and strains carrying the *mecA* gene (38%); most of these isolates carried SCC*mec* type IV [30]. Resistance to other classes of antimicrobial agents was observed in both *mecA* gene-positive and -negative isolates. Strains of *S. aureus* and CoNS carried cassettes characteristic of hospital environments, but there was no difference in antimicrobial susceptibility between SCC*mec* type IV and other SCC*mec* types. No MRSA carrying the PVL gene were detected in that study. Isolates carrying the *luk*S-PV and *luk*F-PV genes were MSSA and were responsible for both primary (furuncle and impetigo) and secondary infections [30].

These results agree with a study conducted by the same group, but involving patients with chronic wounds attending basic health units in the city of Botucatu, Brazil [32]. The prevalence of MRSA isolated from the wounds of 107 patients seen at 18 basic health units was 14.3%. Only SCC*mec* types II and IV were present and the PVL gene was not detected in the isolates studied [32]. These results differ from those reported in other studies, in which these microorganisms are resistant to few non-beta-lactam antibiotics because they frequently carry the PVL genes [33]. An interesting finding obtained in the above study was the isolation of MRSA with the same multidrug-resistant antimicrobial profile from patients of distant basic health units.

Although PFGE is an adequate technique for the study of outbreaks, it is inadequate for long-term studies or global epidemiological studies. In addition to limitations of the interpretation criteria related to short time periods and restricted geographic areas, PFGE presents problems of reproducibility. Different results are obtained when the technique is carried out in different laboratories, even when standardized conditions are used [34]. Studies investigating the clonality of *S. aureus* are frequently complemented by a technique called multilocus sequence typing (MLST). The method is based on the analysis of sequences of seven housekeeping genes of the microorganism, in which different sequences correspond to different alleles of each gene and to a given sequence type (ST). The results of MLST are entered into a database at http://www.mlst.net, which permits comparisons between *S. aureus* sequences described in different parts of the world.

The global clones were classified by MLST as ST-239 (BEC), ST-5 (pediatric and New York/Japan), and ST-247 (Iberian). In a study conducted at São Paulo Hospital, Brazil, MLST of SCC*mec* IV *S. aureus* showed that the isolates belonged to clonal complex 5 (CC5) and carried a new allele, resulting in a new clone (ST-1776) that is emerging in community-associated and nosocomial infections in Brazil [35]. A recent population-based study conducted in the interior of the same state (Botucatu), located 220 km from São Paulo city [36], also detected this clone in healthy individuals, as well as the presence of a new clone (ST-2594) that also belongs to CC5. Recently, our research group published a case report of CA-MRSA infection in a young boy from a small town near Botucatu (Bofete, about 9,000 inhabitants), who had no history of exposure to healthcare services or recent travel. The patient presented with trauma after a soccer game and developed cellulitis, local abscess, pneumonia and severe sepsis. An SCC*mec* type IV MRSA was isolated from blood and tracheal aspirate cultures and was assigned to CC5 (ST-5) [37].

In 2002, Enright et al. [38] analyzed an international collection of nosocomial and community-associated *S. aureus* isolates by SCC*mec* typing and MLST. The results provided new data about clonal groups of *S. aureus*. Different CCs consisting of *S. aureus* isolates with the same or related STs (sharing at least five loci) were identified. Five large MRSA CCs were included: CC8, CC5, CC30, CC45, and CC22. Curiously, the comparison of nucleotide sequences of MRSA and MSSA belonging to CC5 permitted the establishment of a common ancestor of MSSA. Different MRSA isolates would have descended from this MSSA ancestor by independent acquisition of SCC*mec* types.

The origin of SCC*mec* is unknown. According to Wu et al. [39], *S. sciuri* may carry the ancestor of PBP2a, since a PBP found in *S. sciuri* showed 87.8% amino acid sequence similarity to PBP2a. However, *S. sciuri* continues to be susceptible to methicillin because the gene is not expressed. Recently, Hiramatisu et al. [40] suggested an animal-related *Staphylococcus* species, *S. fleurettii*, to be the most likely origin of the *mecA* gene. *Staphylococcus fleurettii* is resistant to methicillin *in vitro* and shows the highest nucleotide homology of the *mecA* gene (99.8%) with *S. aureus* N315, thus indicating *S. fleurettii* instead of *S. sciuri* as a source of the *mecA* gene. Interestingly, the *mecA* gene in *S. fleurettii* is found on the chromosome where it is associated with genes essential for growth, but not with SCC*mec*. Surprisingly, analysis of the *S. fleurettii* genome around the *mecA* gene by PCR-based sequencing revealed a structure that is identical to the class A *mec* complex found in SCC*mec*. However, the *mvaS* gene is intact and not truncated by the insertion sequence IS431*mec* found in MRSA N315. Moreover, the intact *mvaS* gene was followed by the *mvaAC* genes, constituting the mevalonate pathway (six genes essential for the growth of *Staphylococcus*).

Therefore, the study of Hiramatisu et al. [40] indicates that *S. fleurettii* had inherited the *mecA* gene from its ancestor by vertical transmission and had not acquired it recently. The formation of SCC*mec* seems to have occurred by a combination of the two following genetic components: the *mec* gene complex derived from *S. fleurettii* and an SCC element without the *mecA* gene. IS431*mec* may have been involved in the excision of the *mec* complex (*mecA* gene) from the chromosome of *S. fleurettii*, since IS431 is part of many composite transposons. Therefore, since *S. fleurettii* is intrinsically resistant to beta-lactam antibiotics, other methicillin-sensitive *Staphylococcus* species of animal origin coexisting with *S. fleurettii* would be the most likely candidates for SCC*mec* formation [40].

There is still no consensus whether transmission of SCC*mec* occurs only rarely in *S. aureus*, with a limited number of MRSA clones being disseminated globally, or whether transmission of SCC*mec* occurs several times in different strains. However, these new data permit to suggest that SCC*mec* is transmissible in a greater order of magnitude than initially believed, that MRSA clones are emerging on different occasions in different places, and that their geographic dispersal is limited. Furthermore, Wielders et al. [41] isolated an epidemic MSSA strain and subsequently an isogenic MRSA strain from a newborn that had never been in contact with MRSA. The *mecA* gene was identical to that of an *S. epidermidis* strain isolated from this infant. It was suggested that the MRSA strain had originated *in vivo* through horizontal transfer of the *mecA* gene between the two staphylococcal species [41]. These observations support the idea of horizontal transfer of the *mecA* gene and the importance of CoNS as reservoirs of resistance genes that can be transferred to more pathogenic *S. aureus*.

Another commonly used typing method is *spa* typing developed by Frénay et al. [42], which determines sequence variation in the polymorphic X region of protein A of *S. aureus*. In this method, the protein A (*spa*) gene is amplified by PCR and sequenced for analysis of the polymorphic X region or of simple sequence repeats (SSR). The SSR region of the *spa* gene is subject to spontaneous mutations, as well as the loss or gain of repeats. Alphanumeric codes are attributed to these repeats and the *spa* type is deduced from the order of specific repeats [43]. Since *spa* typing involves the sequencing of only one gene, this method has significant advantages in terms of speed, ease of use, standardization, interpretation, interlaboratory comparison and reproducibility when compared to MLST and other techniques such as PFGE [44]. *Spa* typing can be used to study the molecular evolution of MRSA as well as nosocomial outbreaks. A universal nomenclature and public access to the *spa* typing data are guaranteed by the SeqNet.org initiative (www.seqnet.org), which runs the central *spa* server (http://spaserver.ridom.de) on which *spa* typing data are synchronized [45].

Despite the relatively small number of studies investigating the molecular epidemiology of MRSA clones in Brazil, it became clear that different clones are circulating in the country and that these clones differ in their virulence and antimicrobial resistance profile. Characterization of these clones is important to formulate adequate local therapeutic strategies. For example, a more complete knowledge of the clones circulating in a given region can be used to evaluate the relationship between clonal types, signs and symptoms of disease, choice of antimicrobial agents, and clinical outcomes. It is worth noting that more virulent clones start to emerge more frequently both in hospitals and in the community, and there is evidence that virulence factors can be transferred between nosocomial and community-associated clones by recombination. These variable patterns have significant implications for clinical practice. Thus, regional molecular epidemiology programs are needed for the accurate identification and characterization of the MRSA clones circulating in Brazil in order to choose the most appropriate empirical antimicrobial treatment.

REFERENCES

[1] Lindsay, JA. *Staphylococcus*: Molecular Genetics". Caister Academic Press. 1st ed. United Kingdom, 2008.
[2] Tenover, FC; Arbeit, RD; Goering, RV; Mickelsen, PA; Murray, BE; Persing, DH; Swaminathan, B. Interpreting chromosomal DNA restriction patterns produced by pulsed-field gel electrophoresis: criteria for bacterial strain typing. *J Clin Microbiol.*, 1995, 33(9), 2233 – 9.
[3] Chambers, HF; Deleo, FR. Waves of resistance: *Staphylococcus aureus* in the antibiotic era. *Nat Rev Microbiol.*, 2009, 7(9), 629-41.
[4] Sader, HS; Pignatari, AC; Hollis, RJ; Jones, RN. Evaluation of interhospital spread of methicillin-resistant *Staphylococcus aureus* in Sao Paulo, Brazil, using pulsed-field gel electrophoresis of chromosomal DNA. *Infect Control Hosp Epidemiol.*, 1994, 15(5), 320-3.
[5] Gomes, AR; Sanches, IS; Aires, de Sousa, M; Castañeda, E; de Lencastre, H. Molecular epidemiology of methicillin-resistant *Staphylococcus aureus* in Colombian hospitals: dominance of a single unique multidrug-resistant clone. *Microb Drug Resist.*, 2001, 7(1), 23-32.
[6] Aires, De, Sousa, M; Miragaia, M; Sanches, IS; Avila, S; Adamson, I; Casagrande, ST; et al. Three-year assessment of methicillin-resistant *Staphylococcus aureus* clones in Latin America from 1996 to 1998. *J Clin Microbiol.*, 2001, 39(6), 2197-205.0
[7] Oliveira, GA; Faria, JB; Levy, CE; Mamizuka, EM. Characterization of the Brazilian endemic clone of methicillin-resistant *Staphylococcus aureus* (MRSA) from hospitals throughout Brazil. *Braz J Infect Dis.*, 2001, 5(4), 163-70.

[8] Teixeira, LA; Resende, CA; Ormonde, LR; Rosenbaum, R; Figueiredo, AM; de Lencastre, H; et al. Geographic spread of epidemic multiresistant *Staphylococcus aureus* clone in Brazil. *J Clin Microbiol.*, 1995, 33(9), 2400-4.

[9] Wertheim, HF; Melles, DC; Vos, MC; van Leeuwen, W; van Belkum, A; Verbrugh, HA; et al. The role of nasal carriage in *Staphylococcus aureus* infections. *Lancet Infect Dis.*, 2005, 5(12), 751-62.

[10] Hardy, KJ; Oppenheim, BA; Gossain, S; Gao, F; Hawkey, PM. A study of the relationship between environmental contamination with methicillin-resistant *Staphylococcus aureus* (MRSA) and patients' acquisition of MRSA. *Infect Control Hosp Epidemiol.*, 2006, 27(2), 127-32.

[11] Finch, R. Gram-positive infections: lessons learnt and novel solutions. *Clinical Microbiology and Infection.*, 2006, 12 (Suppl. 8), 3–8.

[12] Papia, G; Louie, M; Tralla, A; Johnson, C; Collins, V; Simor, AE. Screening high-risk patients for methicillin-resistant *Staphylococcus aureus* on admission to the hospital: is it cost effective? *Infect Control Hosp Epidemiol.*, 1999, 20(7), 473-7.

[13] Wernitz, MH; Swidsinski, S; Weist, K; Sohr, D; Witte, W; Franke, KP; et al. Effectiveness of a hospital-wide selective screening programme for methicillin-resistant *Staphylococcus aureus* (MRSA) carriers at hospital admission to prevent hospital-acquired MRSA infections. *Clin Microbiol Infect.*, 2005, 11(6), 457-65.

[14] Muto, CA; Jernigan, JA; Ostrowsky, BE; Richet, HM; Jarvis, WR; Boyce, JM; et al. SHEA guideline for preventing nosocomial transmission of multidrug-resistant strains of *Staphylococcus aureus* and *Enterococcus*. *Infect Control Hosp Epidemiol.*, 2003, 24(5), 362-86.

[15] Mombach, P; Machado, AB; Reiter, KC; Paiva, RM; Barth, AL. Distribution of staphylococcal cassette chromosome *mec* (SCC*mec*) types I, II, III and IV in coagulase-negative *staphylococci* from patients attending a tertiary hospital in southern Brazil. *J Med Microbiol.*, 2007, 56(Pt 10), 1328-33.

[16] Bratu, S; Landman, D; Gupta, J; Trehan, M; Panwar, M; Quale, J. A population-based study examining the emergence of community-associated methicillin-resistant *Staphylococcus aureus* USA300 in New York City. *Ann Clin Microbiol Antimicrob.*, 2006, 5, 29.

[17] Ito, T; Katayama, Y; Asada, K; Mori, N; Tsutsumimoto, K; Tiensasitorn, C; et al. Structural comparison of three types of staphylococcal cassette chromosome *mec* integrated in the chromosome

in methicillin-resistant *Staphylococcus aureus*. *Antimicrob Agents Chemother.*, 2001, 45(5), 1323-36.
[18] Bartlett, J. Community-Acquired Methicillin-Resistant *Staphylococcus aureus* and Other Drug-Resistant Pathogens CME. Infectious Diseases Society of America. Annual Meeting 2004.
[19] Dominguez, TJ. It's not a spider bite, it's community-acquired methicillin-resistant *Staphylococcus aureus*. *J Am Board Fam Pract.*, 2004, 17(3), 220-6.
[20] Fitzpatrick, F; Humphreys, H; O'Gara, JP. Evidence for *icaADBC*-independent biofilm development mechanism in methicillin-resistant *Staphylococcus aureus* clinical isolates. *J Clin Microbiol.*, 2005, 43(4), 1973-6.
[21] Rodrigues, MVP; Fortaleza, CMCB; Souza, CSM; Teixeira, NB; Cunha, MLRS. Genetic determinants of methicillin resistance and virulence among *Staphylococcus aureus* isolates recovered from clinical and surveillance cultures in a Brazilian teaching hospital. *ISRN Microb.*, 2012, 1-4.
[22] Rodrigues, MVP; Fortaleza, CMCB; Riboli, DFM; Rocha, RS; Rocha, C; Cunha MLRS. Molecular epidemiology of methicillin-resistant *Staphylococcus aureus* in a burn unit from Brazil. *Burns.*, 2013, 39, 1242-9.
[23] Martins, A; Riboli, DF; Pereira, VC;Cunha, MLRS. Molecular characterization of methicillin-resistant *Staphylococcus aureus* isolated from a Brazilian university hospital. *Braz J Infect Dis.*, 2013. in press http://dx.doi.org/10.1016/j.bjid.2013.11.003
[24] Inoue, F. Caracterização do perfil epidemiológico e molecular de *Staphylococcus aureus* resistentes à oxacilina isolados de hemoculturas de pacientes admitidos no Hospital São Paulo [Dissertation]. São Paulo: Universidade Federal de São Paulo – UNIFESP, Brasil, 2008.
[25] Tomita, ES. Diversidade genética em *Staphylococcus aureus* resistentes à oxacilina em isolados de pacientes pediátricos do Hospital das Clínicas de Botucatu. [Course Completion Work]. Botucatu: Universidade Estadual Paulista Júlio de Mesquita Filho, Brasil, 2011.
[26] Robinson, DA; Enright, MC. Evolutionary models of the emergence of methicillin-resistant *Staphylococcus aureus*. *Antimicrob Agents Chemother.*, 2003, 47(12), 3926-34.
[27] Gardella, N; Murzicato, S; Di, Gregorio, S; Cuirolo, A; Desse, J; Crudo, F; et al. Prevalence and characterization of methicillin-resistant

Staphylococcus aureus among healthy children in a city of Argentina. *Infect Genet Evol.*, 2011, 11(5), 1066-71.

[28] Klevens, RM; Morrison, MA; Nadle, J; Petit, S; Gershman, K; Ray, S; et al. Invasive methicillin-resistant *Staphylococcus aureus* infections in the United States. *JAMA.*, 2007, 298(15), 1763-71.

[29] David, MZ; Siegel, JD; Chambers, HF; Daum, RS. Determining whether methicillin-resistant *Staphylococcus aureus* is associated with health care. *JAMA.*, 2008, 299(5), 519, author reply -20.

[30] Bonesso; MF.Determinação da Virulência e da Resistência Antimicrobiana em *Staphylococcus* spp. Isolados de Pacientes do Serviço de Dermatologia do Hospital das Clínicas da Faculdade de Medicina de Botucatu, SP [Dissertation]. Botucatu: Universidade Estadual Paulista Júlio de Mesquita Filho, Brasil, 2011.

[31] Lu, PL; Chin, LC; Peng, CF; Chiang, YH; Chen, TP; Ma, L; et al. Risk factors and molecular analysis of community methicillin-resistant *Staphylococcus aureus* carriage. *J Clin Microbiol.*, 2005, 43(1), 132-9.

[32] Franchi; EPLP. Caracterização da Resistência à oxacilina em *Staphylococcus aureus* isolados de feridas de pacientes atendidos em unidades básicas de saúde da cidade de Botucatu, SP [Dissertation]. Botucatu: Universidade Estadual Paulista Júlio de Mesquita Filho, Brasil, 2011.

[33] David, MZ; Daum, RS. Community-associated methicillin-resistant *Staphylococcus aureus*: epidemiology and clinical consequences of an emerging epidemic. *Clin Microbiol Rev.*, 2010, 23(3), 616-87.

[34] van Belkum, A; van Leeuwen, W; Kaufmann, ME; Cookson, B; Forey, F; Etienne, J; et al. Assessment of resolution and intercenter reproducibility of results of genotyping *Staphylococcus aureus* by pulsed-field gel electrophoresis of *Sma*I macrorestriction fragments: a multicenter study. *J Clin Microbiol.*, 1998, 36(6), 1653-9.

[35] Carmo, MS; Inoue, F; Andrade, SS; Paschoal, L; Silva, FM; Oliveira, VG; et al. New multilocus sequence typing of MRSA in São Paulo, Brazil. *Braz J Med Biol Res.*, 2011, 44(10), 1013-7.

[36] Pires, FV; Cunha, MLRS; Abraão, LM; Faccioli-Martins, PY; Camargo, CH; Fortaleza, CMCB. Nasal carriage of *Staphylococcus aureus* in Botucatu, Brazil: a population-based survey. *Plos One*, 2014, 9(3): e92537.

[37] Camargo, CH; Cunha, MLRS; Bonesso, MF; Cunha, FP; Barbosa, NA; Fortaleza, CMCB. Systemic CA-MRSA infection following trauma

during soccer match in inner Brazil: clinical and molecular characterization. *Diagn Microbiol Infect Dis.*, 2013, 76(3), 372-4.

[38] Enright, MC; Robinson, DA; Randle, G; Feil, EJ; Grundmann, H; Spratt, BG. The evolutionary history of methicillin-resistant *Staphylococcus aureus* (MRSA). *Proc Natl Acad Sci U S A.*, 2002, 99(11), 7687-92.

[39] Wu, SW; de, Lencastre, H; Tomasz, A. Recruitment of the *mecA* gene homologue S*taphylococcus sciuri* into a resistance determinant and expression of the resistant phenotype in *Staphylococcus aureus. J Bacteriol*, 2001, 183(8), 2417-24.

[40] Hiramatsu, K; Tsubakishita, S; Sasaki, T; Takeuchi, F; Morimoto, Y; Katayama, Y; et al. Genomic basis for methicillin resistance in *Staphylococcus aureus. Infect Chemother*, 2013, 45(2), 117-36.

[41] Wielders, CL; Vriens, MR; Brisse, S; de, Graaf-Miltenburg, LA; Troelstra, A; Fleer, A; Schmitz, FJ; Verhoef, J; Fluit, AC. In-vivo transfer of *mecA* DNA to *Staphylococcus aureus. Lancet.*, 2001, 357(9269), 1674-5.

[42] Frénay, HM; Bunschoten, AE; Schouls, LM; van, Leeuwen, WJ; Vandenbroucke-Grauls, CM; Verhoef, J; et al. Molecular typing of methicillin-resistant *Staphylococcus aureus* on the basis of protein A gene polymorphism. *EurJ Clin Microbiol Infect* Dis., 1996, 15(1), 60-4.

[43] Harmsen, D; Claus, H; Witte, W; Rothgänger, J; Turnwald, D; Vogel, U. Typing of methicillin-resistant *Staphylococcus aureus* in a university hospital setting by using novel software for spa repeat determination and database management. *J Clin Microbiol.*, 2003, 41(12), 5442-8.

[44] Koreen, L; Ramaswamy, SV; Graviss, EA; Naidich, S; Musser, JM; Kreiswirth BN. *spa* typing method for discriminating among *Staphylococcus aureus* isolates: implications for use of a single marker to detect genetic micro- and macrovariation. *J Clin Microbiol.*, 2004, 42(2), 792-9.

[45] Friedrich, AW; Witte, W; Harmsen, D; de Lencastre, H; Hryniewicz, W; Scheres, J; et al. SeqNet.org: a European laboratory network for sequence-based typing of microbial pathogens. *Euro Surveill.*, 2006, 11(1), E060112.4.

AUTHOR'S CONTACT INFORMATION

Maria de Lourdes Ribeiro de Souza da Cunha,
Departamento de Microbiologia e Imunologia
Instituto de Biociências- UNESP-
Rubião Júnior - Botucatu- SP – Brasil
Caixa Postal 510- CEP 18618-970
Email: cunhamlr@ibb.unesp.br

INDEX

A

access, 22, 24, 27, 45, 102
acid, 1, 55, 77
adaptation(s), 2, 9
adhesion, xii, 45, 46, 51, 53, 54, 55, 70
adolescents, 34
aerobic bacteria, 5
aetiology, 34
agar, xi, 25, 27, 29, 51, 84, 85, 89
age, 13, 14
agglutination, xiii, 65, 75, 92
agglutination test, 92
agonist, 70
alanine, 88
allele, 100
amino, 47, 71, 101
amino acid(s), 47, 71, 101
aminoglycosides, 83, 86
amputation, 87
antibiotic, xvi, 11, 23, 37, 52, 81, 86, 88, 96, 103
antibiotic resistance, 96
antibody, 74
antigen, 15, 58
Antimicrobial Resistance, 1, iii, vii, 81
antimicrobial therapy, xvii
apnea, 10, 25
apoptosis, 63
Argentina, 106

aspartic acid, 64
aspirate, 101
assessment, xvi, 25, 103
attachment, 51
autoimmunity, 71
autolysis, 69, 91

B

bacteremia, 9, 17, 21, 23, 25, 28, 34, 35, 43, 54, 56, 75, 97
bacteria, xvi, xvii, 1, 2, 3, 9, 14, 19, 23, 24, 28, 31, 33, 44, 45, 52, 53, 54, 69, 70, 71, 82, 84, 90, 96
bacterial infection, 59
bacterial strains, 40
bacteriophage, 63, 69
bacterium, 2, 10, 13, 70
barriers, 21
base, 77
biofilm, xi, 48, 54, 55, 56
biological activity, 62
biomaterials, 46
biosynthesis, 46, 47
biotechnology, 58
biotic, 45
birds, 2
birth weight, 9, 10, 11, 12, 22, 24
births, 12

blood, xvi, 9, 10, 11, 17, 18, 22, 25, 26, 27, 28, 29, 30, 31, 34, 35, 37, 48, 62, 81, 85, 86, 92, 101
blood cultures, 9, 10, 11, 17, 18, 25, 26, 28, 29, 30, 31, 34, 35, 37, 48, 81, 85, 86, 92
bloodstream, xi, 8, 15, 16, 17, 21, 22, 23, 24, 32, 33, 34, 88, 90, 93
body fluid, 46
bradycardia, 10
Brazil, xv, xvi, 8, 11, 22, 32, 48, 65, 77, 81, 85, 88, 89, 90, 93, 96, 98, 99, 100, 102, 103, 104, 105, 106, 107
breakdown, 21
Britain, 19
burn, 98, 105

C

cadmium, 83
CAM, 17, 56
cancer, 21
candidates, 102
capillary, 65
carbohydrate(s), 39, 55
catheter, xi, 10, 12, 21, 22, 23, 24, 25, 26, 27, 28, 29, 32, 33, 34, 35, 48, 54, 55, 57, 87, 97
cattle, 51, 77
CDC, xi, 8, 10, 17, 22, 24, 33, 86, 87, 88, 89, 90, 92, 99
cDNA, xi, 68
cell death, 62, 73
cell surface, 45, 46, 53, 63
cellulitis, 100
cerebrospinal fluid, 10
children, 11, 12, 34, 106
China, 81
chromosome, xiii, 63, 71, 82, 83, 90, 97, 101, 104
circulation, 99
city, 77, 104
classes, 68, 83, 89, 91, 99
classification, xvi, 1, 8, 37, 39, 99
cleaning, 96
clinical diagnosis, 7
clinical presentation, 10, 12
clinical syndrome, 71
clinical trials, 27
clonality, 100
clone, 90, 95, 96, 98, 99, 100, 103, 104
clustering, 28, 31
clusters, 31
Coagulase-Negative Staphylococci, 1, iii, vii, 7
coccus, 1
coding, 75
collaboration, xv, xvi
collagen, 46
Colombia, 88
colonization, 9, 10, 17, 22, 23, 24, 25, 27, 31, 33, 71, 93, 96, 97
color, 49
commercial, 38, 39, 43, 65
communication, 52, 70
community, xii, xvii, 2, 7, 14, 19, 63, 71, 75, 79, 81, 93, 97, 98, 99, 100, 101, 103, 104, 105, 106
complementary DNA, 68
compliance, 88
complications, 12, 34
composition, 1, 73, 77
computer, 66
Congo, xi, 48, 51
consensus, 24, 102
contaminant, 11, 14
contamination, 9, 10, 11, 18, 21, 22, 24, 27, 37, 42
control measures, 95
correlation, 24, 25
cost, 27, 28, 38, 40, 41, 61, 96, 97, 104
CT, 17, 56
culture, xvi, 10, 25, 26, 27, 28, 29, 34, 56, 65, 66, 68, 70, 71, 89
culture media, 71
culture medium, 68
cytokines, 71
cytotoxicity, 71

D

database, 40, 100, 107
database management, 107
deacetylation, 47
death rate, 21, 24
decay, 61
degradation, 52, 58, 70
Delta, 72
derivatives, 22
detection, xv, xvi, 26, 31, 42, 48, 50, 57, 65, 67, 68, 70, 72, 75, 77, 84, 85, 86, 88, 89, 90, 91, 98
diabetes, 13, 87
diagnostic criteria, 25
dialysis, xi, xv, xvi, 12, 13, 18, 48, 62, 73, 87, 97
diarrhea, 62
diffusion, 26, 29, 30, 65, 85, 89, 91
digestion, 98
discrimination, 40, 42
diseases, ix, 2, 7, 8, 11, 14, 17, 38, 53, 62, 97
dispersion, 52, 71
distilled water, 27
distress, 10
distribution, 75, 97
divergence, 68
diversification, xvii
diversity, 16, 65
DNA, xi, xii, xiii, 28, 30, 35, 36, 41, 44, 51, 58, 61, 63, 67, 68, 91, 98, 103, 107
dominance, 103
drainage, 87
drawing, v
drugs, 2, 3, 31, 45, 54, 82, 90, 97

E

ecology, 57
editors, 3, 4, 5, 7, 15, 17, 42, 78, 90
electrical fields, 98
electrophoresis, xiii, 96, 98, 103, 106
ELISA, xii, 65

elucidation, 28, 32, 69
employees, 89
emulsions, 23
encoding, 61, 64, 66, 67, 68, 69, 74, 75, 82, 86, 98
endocarditis, 4, 7, 10, 63, 91, 97
England, xvi, 12, 18, 33, 73
enteritis, 62
environment(s), 24, 31, 45, 50, 52, 54, 69, 70, 71, 96, 97, 98, 99
environmental conditions, 52
environmental contamination, 104
environmental factors, 69
enzymatic activity, 46
enzyme(s), xv, 2, 13, 46, 53, 63, 65, 67, 68, 81
enzyme-linked immunosorbent assay, 65
epidemic, 95, 96, 102, 104, 106
epidemiology, ix, xv, xvii, 8, 35, 37, 95, 99, 102, 103, 105, 106
Epidemiology of Methicillin-Resistant *Staphylococcus* spp., 95
epithelial cells, 63
epithelium, 9
equipment, 98
erythrocytes, 62, 63
ester, 46
etiology, 8, 12, 13, 14, 22
Europe, 96
evidence, 51, 63, 69, 103
evolution, 9, 10, 12, 13, 16, 21, 57, 62, 65, 96, 99, 102
excision, 82, 101
execution, ix
exotoxins, 53, 98
exposure, 21, 54, 63, 99, 100
external environment, 13
extracellular matrix, 46
extraction, 68

F

false negative, 85
families, 98
fermentation, 1, 38, 39

fetus, 17
fever, 25
fibrin, 23, 33
fibrinogen, 23, 33, 46
financial, ix
financial support, ix
Finland, 15
fitness, 61
flora, 2
fluid, 22, 88
food, 2, 4, 10, 38, 52, 65, 66, 74, 77, 87
food poisoning, 2, 65, 66, 74, 77
formation, xvi, 31, 45, 46, 47, 48, 49, 50, 51, 52, 54, 56, 57, 59, 63, 71, 73, 82, 101
fragments, 98, 106
France, 65, 76
fungal infection, 24
fungi, 14, 23, 24
furuncle, 100

G

gastrointestinal tract, 64
gel, xiii, 40, 65, 96, 98, 103, 106
gene regulation, xvi, 9, 13
genes, 28, 33, 46, 48, 49, 50, 52, 53, 54, 57, 61, 63, 64, 66, 67, 68, 69, 70, 71, 72, 75, 76, 77, 82, 86, 89, 98, 100, 101, 102
genetic components, 101
genetic marker, 35
genetics, 1, 57
genome, 57, 69, 74, 75, 98, 101
genotype, 67
genotyping, 106
genus, xv, xvi, 1, 2, 7, 8, 29, 31
Germany, 7, 15
gestation, 12
glucose, 10, 24
glutamic acid, 64
God, v, ix
granules, 64
growth, 9, 10, 23, 25, 26, 27, 38, 39, 46, 53, 54, 70, 91, 101
guidelines, 88, 90, 97

H

habitats, 3
healing, 59
health, 14, 17, 33, 88, 97, 100, 106
health care, 17, 33, 106
heart valves, 45
heavy metals, 82, 83
hemodialysis, 22
hepatic encephalopathy, 87
heterogeneity, 85
high-risk populations, 21
histidine, 58
history, 5, 99, 100, 107
HM, 43, 104, 107
hospitalization, 12, 24, 96
host, 9, 13, 16, 23, 45, 46, 52, 53, 55, 57, 63, 70, 71, 98
hub, 22, 35
human, 1, 2, 4, 5, 42, 62, 71, 73, 79, 96
human brain, 5
human neutrophils, 71
human skin, 3, 5
husband, v
hybridization, 67, 91
hydrolysis, 63
hydrophobicity, 45, 72
hygiene, 96
hypertension, 87
hypothermia, 25
hypothesis, 13

I

ID, xii, 38, 39, 43
ideal, 68
identification, xii, xv, xvi, 4, 8, 11, 13, 28, 37, 38, 39, 40, 41, 42, 43, 44, 66, 67, 68, 76, 96, 97, 103
immune defense, 45
immune response, 13
immune system, 23
immunocompromised, 2, 3
impetigo, 100

implants, 9, 45
in vitro, 58, 63, 72, 86, 101
in vivo, 33, 59, 63, 64, 71, 102
incidence, 7, 9, 12, 14, 22, 24
incubation time, 38, 40, 84
individuals, 2, 9, 24, 48, 50, 89, 99, 100
inducer, 53
infants, 9, 10, 11, 12, 18, 22, 24, 56
infection, xi, xii, xiv, 3, 8, 9, 10, 11, 12, 14, 15, 17, 18, 19, 21, 22, 24, 25, 26, 31, 32, 33, 34, 38, 43, 48, 54, 55, 56, 59, 62, 69, 70, 71, 79, 87, 88, 89, 93, 95, 96, 97, 98, 100, 106
inflammation, 24, 63, 71
inguinal, 2
inhibitor, 58, 59
initiation, xv
injury, 63, 73
insertion, 21, 22, 24, 34, 97, 101
integration, 82
integrity, 82
intensive care unit, xii, 9, 17, 22, 33, 35, 43, 86
interference, 9, 65, 78
islands, 61, 65, 74
isolation, 10, 24, 26, 27, 43, 88, 96, 97, 99, 100
issues, 93

J

Japan, 86, 95, 100
joints, 45

K

kidney, 7, 87
kidney stones, 87
kill, 23, 54
Kuwait, 77

L

Latin America, 96, 103

lead, 12, 40
lecithin, 62
lesions, 7
lethargy, 10
leukocytes, 63
life cycle, 71
light, 1
linen, 98
lipases, 61
localization, 47
loci, 40, 101
locus, xi, xii, 46, 53, 69, 70, 78
lumen, 27
lupus, 87
lysis, 51, 63

M

macrophages, 23, 63
magnitude, 102
majority, 61
maltose, 38
mammals, 2
man, v
management, 55, 91, 96
mannitol, 38
masking, 68
mastitis, 51, 76
materials, 42, 48
matrix, xii, 23, 46, 73
matter, 54
MB, 15, 16, 32
mechanical ventilation, 10, 11
media, 2
medical, 3, 7, 43, 53, 54, 57, 97
medication, 22
membranes, 72
meningitis, 7, 10, 15
mercury, 83
messenger RNA, 68
metabolism, 13
metabolites, 65
methodology, 8, 21
mice, 15, 63, 79
microbiota, 2, 21, 24, 32, 50

microorganism(s), xvi, xvii, 1, 2, 3, 7, 8, 9, 10, 12, 13, 14, 21, 22, 23, 25, 26, 27, 28, 29, 32, 41, 45, 46, 48, 52, 53, 54, 55, 63, 66, 68, 71, 88, 89, 95, 96, 97, 100
microscope, 1
migration, 21, 96
miniature, 40
models, 64, 105
molecular biology, 1
molecular weight, 19, 44, 64
molecules, xii, 23, 46, 52, 53
monolayer, 46
morbidity, 21, 24, 96
morphology, 1, 37
mortality, 12, 18, 21, 24, 96
MR, 34, 59, 76, 107
mRNA, xii, 53, 68, 69, 71
MRSA, xi, xii, xvi, 5, 63, 64, 73, 75, 78, 79, 82, 84, 85, 86, 88, 89, 90, 91, 93, 95, 96, 97, 98, 99, 100, 101, 102, 103, 104, 106, 107
mucous membrane(s), 2, 9
multiplication, 46
mutagenesis, 53
mutation(s), 44, 102

N

NaCl, 2, 84, 85, 91
nares, 2, 9, 48
necrosis, 63
Neonatal infections, 18
neonates, 17, 32, 72
Netherlands, 97
neutrophils, 23
New York, vii
NH2, 53
nucleoprotein, 63, 73
nucleotide sequence, 75, 101
nursing, 19
nursing home, 19
nutrient(s), 68, 71
nutrition, 9, 11, 22, 23, 24, 33, 34, 70

O

obesity, 87
oligomers, 47
operon, 46, 47, 51, 53, 54, 69, 70, 74
organism, 70, 71
osteomyelitis, 10, 87, 97
outpatient, 14
overproduction, 84
oxygen, 10, 54

P

parallel, xvi
parents, v
pathogenesis, 22, 53, 55, 58, 61, 62, 71, 74
pathogens, 15, 43, 65, 97, 107
PCR, xii, xiii, 28, 29, 30, 31, 35, 36, 40, 41, 44, 48, 49, 66, 67, 68, 69, 72, 76, 77, 86, 101, 102
penicillin, 15, 29, 81, 82, 90, 91
pepsin, 64
peptide(s), xi, xiii, 53, 58, 59, 70, 71, 78, 79
peripheral blood, 26, 27, 28
peritoneum, 62
peritonitis, xv, xvi, 12, 13, 18, 48, 56, 62, 73
Peritonitis, 12, 18
permit, 23, 25, 27, 37, 38, 66, 71, 102
pH, 1, 52, 68, 84
phage, 58, 64, 75, 78, 96
phenol, 70
phenotype(s), 59, 67, 107
Philadelphia, 5, 17, 43, 72
phosphate, 46
phospholipids, 63
phosphorylation, 53, 54, 58
phylum, 1
physicians, xvii
physicochemical properties, 45
plasmid, 65, 75
platelets, 46
PM, 72, 73, 74, 77, 78, 79, 104
pneumonia, 10, 11, 23, 63, 97, 101
policy, 97

polyacrylamide, 40
polymer(s), 5, 55
polymerase, xiii, 28, 40, 44, 48, 66, 76, 77, 86
polymerase chain reaction, xiii, 28, 40, 44, 48, 66, 76, 77, 86
polymorphism(s), 35, 40, 44, 70, 82, 107
polysaccharide, xvi, 23, 45, 46, 47, 50, 56, 57
polystyrene, 48, 49, 50
polyurethane, 24
polyvinyl chloride, 24
poor performance, 40
population, 14, 52, 83, 96, 100, 104, 106
population density, 52
potassium, 2
preparation, 2, 65
preterm infants, 9, 10, 12, 17, 21, 35
prevention, 33, 34, 54
primate, 64
principles, 28
prisons, 98
probability, 13, 48
producers, 48, 49, 66, 67
professionals, 88, 97
prognosis, 81
protection, 70
protein components, 23
proteins, 23, 46, 50, 53, 55, 63, 70, 82, 91
proteolytic enzyme, 64
Pseudomonas aeruginosa, 22, 51, 58, 59
public health, xvii
purification, 55, 65
pus, 1
pyogenic, 7

R

RE, 59
reactions, 29, 31, 65, 66
reactive oxygen, 64
receptors, 23
recognition, 88
recombination, 44, 103
recovery, 15

recurrence, 12
red blood cells, 62
regression, 11, 13
regression analysis, 11
regression model, 13
regulatory systems, 70, 71
reliability, 25, 28, 38, 41
renin, 64
repressor, 53, 69
requirements, 70
researchers, xv, xvii, 9, 62, 63, 66
residues, 46, 47, 58
resistance, xii, xv, xvi, 2, 3, 10, 12, 13, 28, 31, 33, 37, 39, 42, 54, 57, 59, 71, 78, 81, 82, 83, 84, 85, 86, 88, 89, 90, 91, 92, 93, 95, 97, 98, 102, 103, 105, 107
resolution, 13, 48, 62, 106
resources, 38
response, xiv, 69
restriction enzyme, 98
restrictions, 96
reverse transcriptase, 68
RH, 4, 15, 42
rheumatoid arthritis, 87
ribosomal RNA, 1, 40
risk(s), 10, 11, 12, 14, 19, 21, 24, 48, 96, 98, 104
risk factors, 19, 48, 98
RNA, xii, 40, 53, 58, 69
routes, 21
routines, 12
Royal Society, 33
rules, 97

S

salt concentration, 2
samplings, 10
school, 89, 98
sclerosis, 87
SEA, xiii, 64, 65, 66, 67, 68
secrete, 9
secretion, 98
SED, xiii, 65, 68
seed, 1

Index

seeding, 27
sensing, xiii, 52, 53, 54, 58, 59, 70, 71, 78
sensitivity, 19, 25, 26, 28, 39, 40, 41, 49, 65, 66, 85
sepsis, 10, 11, 12, 17, 21, 24, 25, 33, 34, 48, 101
sequencing, 44, 75, 101, 102
serum, 46, 76
services, 99, 100
SES, xiv, 64
sexual activity, 14
sheep, 63
shock, xiv, 74, 76, 77
showing, 25, 26, 49, 85
signal transduction, 70
signals, 78
signs, 10, 11, 25, 103
skin, 2, 3, 9, 16, 21, 22, 24, 45, 63, 64, 97, 98, 99
soccer, 100, 107
sodium, 1
software, 42, 107
solution, 32
somatic cell, 76
SP, xv, xvi, 19, 55, 73, 106
species, xv, xvi, 2, 3, 4, 5, 7, 8, 9, 10, 11, 12, 13, 14, 22, 23, 24, 25, 26, 28, 29, 31, 37, 38, 39, 40, 41, 42, 43, 45, 48, 51, 54, 62, 64, 67, 68, 69, 70, 71, 77, 101, 102
SS, 64, 78, 79, 106
standardization, 25, 68, 102
staphylococci, xi, xv, xvi, 1, 2, 4, 8, 13, 14, 22, 29, 35, 37, 38, 40, 41, 43, 45, 49, 50, 51, 54, 56, 57, 62, 64, 65, 66, 67, 71, 76, 77, 79, 81, 82, 89, 90, 91, 92, 93, 99, 104
Staphylococcus aureus, xi, xii, xiii, xiv, xv, xvi, 2, 5, 14, 16, 18, 19, 22, 33, 35, 53, 57, 58, 59, 72, 73, 74, 75, 76, 77, 78, 82, 83, 87, 89, 90, 91, 92, 93, 95, 103, 104, 105, 106, 107
state, 45, 89, 100
sterile, 27, 50, 98
structure, 46, 71, 101
success rate, 96
sucrose, 38, 42

supervision, xv
supplementation, 84
suppression, 53
surface component, xii, 23, 46
surface properties, 46, 55
surfactant, 79
surveillance, 8, 15, 16, 17, 18, 25, 32, 64, 86, 88, 89, 97, 98, 105
survival, 9, 21, 24, 52
susceptibility, xvi, 26, 29, 34, 37, 39, 41, 54, 62, 82, 84, 86, 88, 89, 90, 91, 92, 99
symptoms, xvi, 2, 10, 61, 103
syndrome, xiv, 74, 76, 77
synthesis, 46, 47, 53, 56, 82, 88, 91

T

T lymphocytes, 71
target, 28, 44, 52, 53, 58, 69
teams, 98
techniques, xvi, 21, 27, 28, 31, 32, 38, 41, 66, 68, 86, 91, 95, 102
teicoplanin, 69, 88, 90
temperature, 1, 52, 63, 68, 84, 91
testing, 26, 29, 41, 86, 89, 91
therapy, 11, 82, 88
thermostability, 62
time periods, 100
tissue, 10, 53, 56, 63, 70, 71, 87
TMC, 18, 73
toxic shock syndrome, 61, 64, 72, 76, 77
toxin, xiii, xiv, xvi, 61, 62, 63, 64, 66, 67, 68, 69, 70, 71, 72, 73, 74, 75, 76, 77, 98
Toxins, vii, 61, 72, 73
transcription, 53, 56, 71, 77
transcripts, 53
transfer RNA, 40
transmission, 9, 16, 31, 35, 89, 96, 98, 102, 104
trauma, 3, 100, 106
treatment, 21, 23, 28, 53, 54, 81, 82, 85, 86, 88, 89, 92, 96, 103
trypsin, 64
tuberculosis, 35

U

UK, 18
ulcer, 87
United, xiv, 8, 10, 21, 78, 86, 87, 88, 92, 103, 106
United Kingdom, 78, 103
United States, xiv, 8, 10, 21, 86, 87, 88, 92, 106
urethral syndrome, 15
urinary tract, 8, 14, 15, 19, 23, 33, 44, 87
urinary tract infection, 8, 19, 23, 33, 44, 87
urine, 10, 14, 42, 76
USA, xiv, 96

V

vagina, 66
vancomycin, 13, 29, 85, 86, 87, 88, 89, 90, 92, 93, 97
variations, 49, 70
vein, 27, 34
venipuncture, 10
vertical transmission, 101
vessels, 21

W

Washington, 4, 5, 42, 90
Western blot, 69
workers, 66, 77
workload, 34
worldwide, xvii, 86, 95, 96
wound infection, 7, 15

Y

yield, 66